# THE
# GREEN NEW DEAL:

# DESCENT
# INTO
# HELL

**Lawrence Newman**

The Green New Deal: Descent Into Hell

Copyright © 2021 by Lawrence Newman

ISBN 978-1-7347100-4-5

Library of Congress Control Number: 2021933510

Printed in the United States of America

First  Edition

Publisher: Silver Millennium Publications, Inc.
            Gold Canyon, Arizona

## *Other books by the author*

The Apostle Islands — America's Wilderness In The Water

Discovering The Apostle Islands

Sailing Adventures In The Apostle Islands

Tales Of A Nautical Novice:
Lessons I Learned Boating In The Great Lakes

Traveler's Guide to Wisconsin's Lake Superior Shore

Don't Leave Your Children's Education to the Educators

365 Outstanding Quotes That Could Change Your Life

1000 Outstanding Quotes You Should Know

120 Quotations That Every Educated Person Should Know

Treasury of "On Target" Humorous Quotes For Public Speakers

Motivational Quotes For Coaches, Captains and Team Leaders

Self-Publishing Your Book: A Nuts and Bolts Approach

The  Climate Change Hoax: Pathway to Socialism

Don't Believe In God?

*Dedicated to all those who question the rush to judgment on man-made climate change and where it is leading us*

# TABLE OF CONTENTS

Chapter 1 :  Introduction

Chapter 2 :  The Texas Tragedy: A Glimpse
Into The Future

Chapter 3:  The "Green New Deal"—The Devil's
In The Details

Chapter 4 :  Dissecting The "Deal"

Chapter 5 :  The Hell of Socialism

Chapter 6:   The Devil's Twin Brother: The Paris
Climate Accord

Chapter 7 :    The Struggle Ahead

Relevant Quotes

Appendix I:   Speech by Governor Nikki Haley

Appendix II:   The Paris Climate Accord

Reference Sources

# CHAPTER 1
# INTRODUCTION

Shortly after President Biden's inauguration in January 2021, events began to move quickly towards the implementation of programs sponsored by the progressive left. Chief among their goals is the passage of the Green New Deal, based on the faulty assumptions underlying man-made climate change.

The end result of the implementation of the programs incorporated in the proposed legislation would be to devastate the economy of the United States, resulting in a life for its citizens that would be nasty, brutish and short—similar to what happened to the citizens of Venezuela.

This book has been written to expose the erroneous assumptions underlying this legislation and the actual goals inherent in the Green New Deal. The ultimate goal of this legislation is to destroy the free-market economy of the United States and turn it into a socialistic, government-run system that destroys all individual initiative.

In accordance with the ultimate goal of the Green New Deal, one of the first actions taken by the new administration in February 2021 was the reentry of the United States into the Paris Climate Accord, which President Trump had taken the United States out of in 2017. This one-sided agreement is detrimental to America's interests in many ways, which will also be discussed.

It is incumbent on the citizens of the United States to fight against the actions of the progressive left by all means necessary, but specifically through the ballot box in order to prevent the implementation of their plans to turn the United States into a socialistic dictatorship.

# CHAPTER 2

## THE TEXAS TRAGEDY:
## A GLIMPSE INTO THE FUTURE

In mid-February 2021 a tremendous cold wave hit the United States from the East Coast down to the Mexican border. The bitter cold and accompanying snow and freezing rain affected all parts of this area but had its most devastating impact on the 29 million citizens of Texas.

The extreme cold caused the state's electricity grid to fail in spectacular fashion resulting in a lack of water and heat for its citizens—and dozens died. Just days before, Bill Magness, chief executive of the comically titled Electric Reliability Council of Texas had assured the state's citizens that the electric grid was ready for the cold wave that was forecast. The intense cold that descended upon the state impacted unprotected gas line valves and water pipes, including those used for cooling the South Texas Nuclear Generating Station. However, its most devasting effect was on the wind turbines which generate almost a quarter of the state's electrical requirements—when the wind is blowing and they aren't incapacitated due to freezing rain. They simply locked up and became useless. And therein lies the inherent falsity of the promises of the Green New Deal.

Here are the facts of the Texas tragedy, according to the Energy Information Administration:

Between 12 AM on February 8 and February 16 wind generated power plunged 93% while coal generated power increased 47% and the gas generated power increased 450%. Unfortunately, the increase in the coal and gas power generation was insufficient to offset the wind power loss of nearly 100%, since wind power had become just a large component of Texas's power base.

Prior to the Texas tragedy the citizens of California experienced the inconvenience of rolling blackouts in previous years as the state deactivated many of its fossil fuel plants and moved to green-based energy production. Paradoxically, in order to alleviate the rolling blackouts California had to resort to buying fossil-fuel generated electricity from other states, which resulted in increased costs to its citizens.

The nation's electrical needs will never be met by "green friendly" generation alone. These methods may have a place in the total picture but in order to have a reliable electrical grid a substantial portion of the country's power needs must be supplied by oil, gas, coal and nuclear. Many of the fuel line valves to the oil and gas generating Texas plants were impacted by the cold.. However, the fuel for coal plants was not impacted by the cold since it was available on site to provide the energy needed to run its generators.

It's an obvious fact that wind and solar power generation does not occur when the wind's not blowing and the sun's not shining. And furthermore, both are impacted by extreme cold, snow and freezing rain. Coal fired plants are not substantially impacted by these weather events. Although supporters of green methods of electrical generation are placing their bets on future breakthroughs in battery capabilities there are no such breakthroughs visible on the horizon.

In an effort to help alleviate the plight of the Texas citizenry the Biden administration authorized the transfer of diesel generators. The thought of dirty diesel fueled generators coming to the rescue of a state which embraced green energy must have been galling to the supporters of the Green New Deal. It's obvious that the harder green activists attempt to get rid of fossil fuels the more we find that we can't live without them.

It's an interesting fact that China recognizes the reliability that electricity generating plants powered by coal provides since they are building substantial numbers of them. A recent report by the Global Energy Monitor indicated that China built over three times as much new coal power capacity In 2020, as all other countries in the world combined. China also initiated planning for the future construction of 73.5 gigawatts of new coal plants—five times as much as the rest of the world combined.-

In addition, China is building nuclear facilities, which Green New Deal advocates abhor. It won't be long before China's electrical grid will be more reliable than our own—if it isn't already. There are no rolling blackouts in China. Millions of our citizens are being sacrificed on the altar of the Green New Deal and will experience the inevitable breakdown of the country's electrical grid in the future. The recent  events in Texas have allowed us a glimpse of things to come.

In response to the Texas calamity,  Governor Greg Abbott urged lawmakers to  pass legislation to make sure the energy grid would never again be inoperable in cold weather.  "What happened this week to our fellow Texans is  completely unacceptable and can never be replicated again."

It will become obvious to Texas lawmakers that  it was an over-reliance on green energy production that was one of the major reasons leading to the fiasco they experienced.  Fossil fuel and nuclear power generation will always be an essential part of an electrical grid if reliability is wanted.

Nature taught Texas and the country this hard lesson.  We should never be in a situation, sitting in the dark, enveloped in frigid cold, without access to drinking water,  wondering how we got there. Implementation of the provisions of the Green New Deal will lead to this result—if we let it.

# CHAPTER 3

## THE "GREEN NEW DEAL"—
## THE DEVIL'S IN THE DETAILS

Shown below is House Resolution 109, the "Green New Deal" that was introduced in the House of Representatives in early 2019. By the end of the 116th Congress the resolution had not come up for a vote in the House. However, since the Republicans controlled the Senate in the 116th Congress it was put up for a vote to see if those Senators who said they supported the socialistic thrust of this climate change agenda would vote for it. The final vote was 57-0 <u>against</u> the resolution with 43 Democrats voting "Present". To their credit four Democrats joined all the Republicans voting against it.

The resolution, in similar form, is likely to be reintroduced in the Democratically controlled House during the next Congress..

A critique of this assault on our common sense follows in Chapter 4.

116TH CONGRESS
1ST SESSION

## HOUSE RESOLUTION 109

Recognizing the duty of the Federal Government to create a Green New Deal.

# IN THE HOUSE OF REPRESENTATIVES

FEBRUARY 7, 2019

Ms. OCASIO-CORTEZ (for herself, Mr. HASTINGS, Ms. TLAIB, Mr. SERRANO, Mrs. CAROLYN B. MALONEY of New York, Mr. VARGAS, Mr. ESPAILLAT, Mr. LYNCH, Ms. VELÁZQUEZ, Mr.BLUMENAUER, Mr. BRENDAN F. BOYLE of Pennsylvania, Mr. CASTRO of Texas, Ms. CLARKE of New York, Ms. JAYAPAL, Mr. KHANNA, Mr. TED LIEU of California, Ms. PRESSLEY, Mr. WELCH, Mr. ENGEL, Mr. NEGUSE, Mr. NADLER, Mr. MCGOVERN, Mr. POCAN, Mr. TAKANO, Ms. NORTON, Mr. RASKIN, Mr. CONNOLLY, Mr. LOWENTHAL, Ms. MATSUI, Mr. THOMPSON of California, Mr. LEVIN of California, Ms. PINGREE, Mr. QUIGLEY, Mr. HUFFMAN, Mrs. WATSON COLEMAN, Mr. GARCÍA of Illinois, Mr. HIGGINS of New York, Ms. HAALAND, Ms. MENG, Mr. CARBAJAL, Mr. CICILLINE, Mr. COHEN, Ms. CLARK of Massachusetts, Ms. JUDY CHU of California, Ms. MUCARSEL-POWELL, Mr. MOULTON, Mr. GRIJALVA, Mr. MEEKS, Mr. SABLAN, Ms. LEE of California, Ms. BONAMICI, Mr. SEAN PATRICK MALONEY of New York, Ms. SCHAKOWSKY, Ms. DELAURO, Mr. LEVIN of Michigan, Ms. MCCOLLUM, Mr. DESAULNIER, Mr. COURTNEY, Mr.LARSON of Connecticut, Ms. ESCOBAR, Mr. SCHIFF, Mr. KEATING, Mr. DEFAZIO, Ms. ESHOO, Mrs. TRAHAN, Mr. GOMEZ, Mr. KENNEDY, and Ms. WATERS) submitted the following resolution; which was referred to the Committee on Energy and Commerce, and in addition to the Committees on Science, Space, and Technology, Education and Labor,

8

Transportation and Infrastructure, Agriculture, Natural Resources, Foreign Affairs, Financial Services, the Judiciary, Ways and Means, and Oversight and Reform, for a period to be subsequently determined by the Speaker, in each case for consideration of such provisions as fall within the jurisdiction of the committee concerned

---

# RESOLUTION

Recognizing the duty of the Federal Government to create a Green New Deal.

Whereas the October 2018 report entitled "Special Report on Global Warming of 1.5 °C" by the Intergovernmental Panel on Climate Change and the November 2018 Fourth National Climate Assessment report found that—

(1) human activity is the dominant cause of observed climate change over the past century;

(2) a changing climate is causing sea levels to rise and an increase in wildfires, severe storms, droughts, and other extreme weather events that threaten human life, healthy communities, and critical infrastructure;

(3) global warming at or above 2 degrees Celsius beyond pre-industrialized levels will cause—

(A) mass migration from the regions most affected by climate change;

(B) more than $500,000,000,000 in lost annual economic output in the United States by the year 2100;

(C) wildfires that, by 2050, will annually burn at least twice as much forest area in the western United States than was typically burned by wildfires in the years preceding 2019;

(D) a loss of more than 99 percent of all coral reefs on Earth;

(E) more than 350,000,000 more people to be exposed globally to deadly heat stress by 2050; and

(F) a risk of damage to $1,000,000,000,000 of public infrastructure and coastal real estate in the United States; and

(4) global temperatures must be kept below 1.5 degrees Celsius above pre-industrialized levels to avoid the most severe impacts of a changing climate, which will require—

(A) global reductions in greenhouse gas emissions from human sources of 40 to 60 percent from 2010 levels by 2030; and

(B) net-zero global emissions by 2050;

Whereas, because the United States has historically been responsible for a disproportionate amount of greenhouse gas emissions, having emitted 20 percent of global greenhouse gas emissions through 2014, and has a high technological capacity, the United States must take a leading role in reducing emissions through economic transformation;

Whereas the United States is currently experiencing several related crises, with—

(1) life expectancy declining while basic needs, such as clean air, clean water, healthy food, and adequate health care, housing, transportation, and education, are inaccessible to a significant portion of the United States population;

(2) a 4-decade trend of wage stagnation, deindustrialization, and antilabor policies that has led to—

(A) hourly wages overall stagnating since the 1970s despite increased worker productivity;

(B) the third-worst level of socioeconomic mobility in the developed world before the Great Recession;

(C) the erosion of the earning and bargaining power of workers in the United States; and

(D) inadequate resources for public sector workers to confront the challenges of climate change at local, State, and Federal levels; and

(3) the greatest income inequality since the 1920s, with—

(A) the top 1 percent of earners accruing 91 percent of gains in the first few years of economic recovery after the Great Recession;

(B) a large racial wealth divide amounting to a difference of 20 times more wealth between the average white family and the average black family; and

(C) a gender earnings gap that results in women earning approximately 80 percent as much as men, at the median;

Whereas climate change, pollution, and environmental destruction have exacerbated systemic racial, regional, social, environmental, and economic injustices (referred to in this preamble as "systemic injustices") by disproportionately affecting indigenous peoples, communities of color, migrant communities, deindustrialized communities, depopulated rural communities, the poor, low-income workers, women, the elderly, the unhoused, people with disabilities, and youth (referred to in this preamble as "frontline and vulnerable communities");

Whereas, climate change constitutes a direct threat to the national security of the United States—

(1) by impacting the economic, environmental, and social stability of countries and communities around the world; and

(2) by acting as a threat multiplier;

Whereas the Federal Government-led mobilizations during World War II and the New Deal created the greatest middle class that the United States has ever seen, but many members of frontline and vulnerable communities were excluded from many of the economic and societal benefits of those mobilizations; and

Whereas the House of Representatives recognizes that a new national, social, industrial, and economic mobilization on a scale not seen since World War II and the New Deal era is a historic opportunity—

(1) to create millions of good, high-wage jobs in the United States;

(2) to provide unprecedented levels of prosperity and economic security for all people of the United States; and

(3) to counteract systemic injustices: Now, therefore, be it

*Resolved,* That it is the sense of the House of Representatives that—

(1) it is the duty of the Federal Government to create a Green New Deal—

(A) to achieve net-zero greenhouse gas emissions through a fair and just transition for all communities and workers;

(B) to create millions of good, high-wage jobs and ensure prosperity and economic security for all people of the United States;

(C) to invest in the infrastructure and industry of the United States to sustainably meet the challenges of the 21st century;

(D) to secure for all people of the United States for generations to come—

(i) clean air and water;

(ii) climate and community resiliency;

(iii) healthy food;

(iv) access to nature; and

(v) a sustainable environment; and

(E) to promote justice and equity by stopping current, preventing future, and repairing historic oppression of indigenous peoples, communities of color, migrant communities, deindustrialized communities, depopulated rural communities, the poor, low-income workers, women, the elderly, the unhoused, people with disabilities, and youth (referred to in this resolution as "frontline and vulnerable communities");

(2) the goals described in subparagraphs (A) through (E) of paragraph (1) (referred to in this resolution as the "Green New Deal goals") should be

accomplished through a 10-year national mobilization (referred to in this resolution as the "Green New Deal mobilization") that will require the following goals and projects—

(A) building resiliency against climate change-related disasters, such as extreme weather, including by leveraging funding and providing investments for community-defined projects and strategies;

(B) repairing and upgrading the infrastructure in the United States, including—

(i) by eliminating pollution and greenhouse gas emissions as much as technologically feasible;

(ii) by guaranteeing universal access to clean water;

(iii) by reducing the risks posed by climate impacts; and

(iv) by ensuring that any infrastructure bill considered by Congress addresses climate change;

(C) meeting 100 percent of the power demand in the United States through clean, renewable, and zero-emission energy sources, including—

(i) by dramatically expanding and upgrading renewable power sources; and

(ii) by deploying new capacity;

(D) building or upgrading to energy-efficient, distributed, and "smart" power grids, and ensuring affordable access to electricity;

(E) upgrading all existing buildings in the United States and building new buildings to achieve maximum energy efficiency, water efficiency, safety, affordability, comfort, and durability, including through electrification;

(F) spurring massive growth in clean manufacturing in the United States and removing pollution and greenhouse gas emissions from manufacturing and industry as much as is technologically feasible, including by expanding renewable energy manufacturing and investing in existing manufacturing and industry;

(G) working collaboratively with farmers and ranchers in the United States to remove pollution and greenhouse gas emissions from the agricultural sector as much as is technologically feasible, including—

(i) by supporting family farming;

(ii) by investing in sustainable farming and land use practices that increase soil health; and

(iii) by building a more sustainable food system that ensures universal access to healthy food;

(H) overhauling transportation systems in the United States to remove pollution and greenhouse gas emissions from the transportation sector as much as

16

is technologically feasible, including through investment in—

(i) zero-emission vehicle infrastructure and manufacturing;

(ii) clean, affordable, and accessible public transit; and

(iii) high-speed rail;

(I) mitigating and managing the long-term adverse health, economic, and other effects of pollution and climate change, including by providing funding for community-defined projects and strategies;

(J) removing greenhouse gases from the atmosphere and reducing pollution by restoring natural ecosystems through proven low-tech solutions that increase soil carbon storage, such as land preservation and afforestation;

(K) restoring and protecting threatened, endangered, and fragile ecosystems through locally appropriate and science-based projects that enhance biodiversity and support climate resiliency;

(L) cleaning up existing hazardous waste and abandoned sites, ensuring economic development and sustainability on those sites;

(M) identifying other emission and pollution sources and creating solutions to remove them; and

17

(N) promoting the international exchange of technology, expertise, products, funding, and services, with the aim of making the United States the international leader on climate action, and to help other countries achieve a Green New Deal;

(3) a Green New Deal must be developed through transparent and inclusive consultation, collaboration, and partnership with frontline and vulnerable communities, labor unions, worker cooperatives, civil society groups, academia, and businesses; and

(4) to achieve the Green New Deal goals and mobilization, a Green New Deal will require the following goals and projects—

(A) providing and leveraging, in a way that ensures that the public receives appropriate ownership stakes and returns on investment, adequate capital (including through community grants, public banks, and other public financing), technical expertise, supporting policies, and other forms of assistance to communities, organizations, Federal, State, and local government agencies, and businesses working on the Green New Deal mobilization;

(B) ensuring that the Federal Government takes into account the complete environmental and social costs and impacts of emissions through—

(i) existing laws;

(ii) new policies and programs; and

(iii) ensuring that frontline and vulnerable communities shall not be adversely affected;

(C) providing resources, training, and high-quality education, including higher education, to all people of the United States, with a focus on frontline and vulnerable communities, so that all people of the United States may be full and equal participants in the Green New Deal mobilization;

(D) making public investments in the research and development of new clean and renewable energy technologies and industries;

(E) directing investments to spur economic development, deepen and diversify industry and business in local and regional economies, and build wealth and community ownership, while prioritizing high-quality job creation and economic, social, and environmental benefits in frontline and vulnerable communities, and deindustrialized communities, that may otherwise struggle with the transition away from greenhouse gas intensive industries;

(F) ensuring the use of democratic and participatory processes that are inclusive of and led by frontline and vulnerable communities and workers to plan, implement, and administer the Green New Deal mobilization at the local level;

(G) ensuring that the Green New Deal mobilization creates high-quality union jobs that pay prevailing wages, hires local workers, offers training and advancement opportunities, and guarantees wage and benefit parity for workers affected by the transition;

(H) guaranteeing a job with a family-sustaining wage, adequate family and medical leave, paid vacations, and retirement security to all people of the United States;

(I) strengthening and protecting the right of all workers to organize, unionize, and collectively bargain free of coercion, intimidation, and harassment;

(J) strengthening and enforcing labor, workplace health and safety, antidiscrimination, and wage and hour standards across all employers, industries, and sectors;

(K) enacting and enforcing trade rules, procurement standards, and border adjustments with strong labor and environmental protections—

(i) to stop the transfer of jobs and pollution overseas; and

(ii) to grow domestic manufacturing in the United States;

(L) ensuring that public lands, waters, and oceans are protected and that eminent domain is not abused;

(M) obtaining the free, prior, and informed consent of indigenous peoples for all decisions that affect indigenous peoples and their traditional territories, honoring all treaties and agreements with indigenous peoples, and protecting and enforcing the sovereignty and land rights of indigenous peoples;

(N) ensuring a commercial environment where every businessperson is free from unfair competition and domination by domestic or international monopolies; and

(O) providing all people of the United States with—

(i) high-quality health care;

(ii) affordable, safe, and adequate housing;

(iii) economic security; and

(iv) clean water, clean air, healthy and affordable food, and access to nature.

# CHAPTER 4

## DISSECTING THE "DEAL"

Before addressing each of its proposals it might be useful to first start with the subject of its implementation cost—estimated at approx.-imately $94 trillion. The supporters of the plan are calling for a vast increase in taxes as a starting point. However, we should keep in mind a famous quote of the late Prime Minister of Britain, Margaret Thatcher, "The problem with socialism is that eventually you run out of other people's money."

Since the implementation of its provisions will destroy our capitalistic economy it's difficult to determine where the additional monies to pay for the completion of the socialistic paradise will come from since it doesn't appear there are any provisions in the legislation for the planting of groves of money trees. If their intention is to start running the Treasury's money presses around the clock our streets we'll quickly begin looking like Venezuela where it takes a wheelbarrow of money to buy a loaf of bread. For the majority of our citizens life will quickly revert to the standard of the Middle Ages where life was "nasty, brutish and short".

Ask the Venezuelans how things turned out for them when they converted from a capitalistic to a socialistic economy.

The opening pages of the House Resolution begins with a series of lies:

- Human activity has <u>not</u> been proven to be the dominant cause of observed climate change in the past century. Where's the proof for that outrageous statement?

- Although there may be a small increase in sea levels where is the proof that this has led to an increase in wildfires and extreme weather events?

- Where is the proof that a couple of degrees of temperature change will cause the dire effects they prophesize? The $500 trillion figure they attribute as the economic loss in the United States is pure fantasy. They cannot back up that figure. Even the UN Panel disagrees with them.

  *"Experts know that the worst-case climate projections show minimal impact on the overall economy. Buried in the Intergovernmental Panel on Climate Change 2014 report is a chart showing that a global temperature rise of 5 degrees Fahrenheit would have a global economic impact of about 3% in 2100, negligibly diminishing*

*projected global growth over that period to 385% from 400%. There are many reasons to be concerned about a changing climate, including disparate impact across industries and regions. But national economic catastrophe isn't one of them. It should concern anyone who supports well-informed public and policy discussions that the report's authors, reviewers and media coverage obscured such an important point.—Steven Koonin*

- The other dire figures they state in their opening statements are unable to be supported by reputable scientists and are pulled from thin air.

Having opened their resolution with a plethora of lies they then propose their solutions:

- A global reduction of 40-60% in carbon dioxide by 2030.

- Net-zero global emissions by 2050

Not only are these reductions impossible for the United States, the resolution specifies these are <u>global</u> goals. Good luck imposing these reductions on the Chinese, Russians, Iranians, North Koreans and the rest of the world. I'm sure the other countries of the world won't spend much time dealing with the demands of our climate change fanatics. Unfortunately, we have to.

The resolution then goes on to list the problems we currently have in the United States, none of which have anything to do with global warming. The Green New Deal is being put forward to correct all perceived problems in our capitalistic society and convert it to socialism, using identity politics as the driving force—rich against poor, black against white, men against women, old against young, urban against rural, the list is endless. It's also obvious that the proponents of this legislation are living in an alternate universe. We do have problems of homelessness, medical coverage and areas of poverty that must be addressed, but the United States is still the "shining beacon on the hill" that many in the world want to come to . We provide the highest standard of living and more freedom for our citizens than any other major country on Earth.

The resolution then goes on to list all the objectives that it wishes to fulfill including clean air and water, healthy food, high wage jobs, access to nature . . . another list that is endless!

What follows are the specific actions the resolution envisages including:

- Attaining 100% emission-free power generation, which means all coal, oil and gas power generating plants would be deactivated. Since nuclear energy is not presented as an acceptable choice that pretty much leaves solar and wind as the major sources. All leading experts in power generation say this objective is absolutely impossible for many reasons, including the lack of sufficient battery resources to store energy when the wind isn't blowing and the sun isn't shining. No known electrical network grid is available to handle a 100% solar/wind array of power generation. The recent Texas tragedy resulting from a dependence on renewables, illustrating this fact, was described in Chapter 2.

- The upgrading of all presently constructed buildings in the United States to make them energy efficient. We can say goodbye to glass and steel structures internally well-lit with natural light that currently comprise the bulk of our major city business districts. Again, we have to ask—where is all the money coming from to retrofit all of these buildings?

- Removing all pollution and carbon dioxide generation from manufacturing plants in the United States. There goes our balance of trade since apparently we will now be importing all manufactured goods from countries who aren't as stupid as we are.

- Removing pollution and greenhouse gas emissions from the agricultural sector. So much for you steak eaters since those methane-emitting cows will have to go. It isn't assured that any meat will remain on future menus since all farm animals generate pollution. The solution offered of eating so-called environmentally preferred, plant-based "meat flavored" products should be a non-starter for those interested in removing the chemicals from the foods they eat. Anyone looked at the ingredients on the labels of those fake meat products lately?

- Overhauling the transportation systems in the United States. All internal combustion cars will be forbidden. It's very doubtful long-range batteries will be developed in the near future that could power a large car or RV across the country in the time that gas power can. The Indianapolis 500 and stock car races will definitely be attractions of the past. It appears long-haul trucking will have to go, since high-speed rail will be replacing trucks and planes. Good luck with that. The building of a high-speed rail network will be the biggest boondoggle of all time if the example set by just one construction attempt is indicative—that of

the proposed high-speed rail project
between Los Angeles and San Francisco.

*"I would suggest that the projections might be a bit optimistic."—Karl Compton, Letters to the Editor in The Wall Street Journal, who pointed out that the plans for the construction of a $40 billion dollar bullet train system from Los Angeles to San Francisco projected 41 million riders per year (40% of breakeven). " . . . that works out to 112,328 riders per day or 4680 per hour. With 150 passengers packed cheek-to-jowl into each car, that's a 31-car train leaving every hour of the day and night."*

Since a round-trip air ticket between San Francisco and Los Angeles currently goes for approximately $200, one has to question how many travelers would want to pony up $400 for a one-way train ticket. Anyone else have a multibillion-dollar boondoggle idea to build a terrorist target in an earthquake prone region?

Note: In early 2019 California Governor Newsom canceled the partially completed bullet train project when its projected cost rose to $77 billion and its completion date was 100 years in the future at its current construction rate.

Since ships and planes will also be on the chopping block we will have to come up with a means of travel to Hawaii, Asia and Europe. Underwater trains, anyone?

Under the Green New Deal economic security from cradle to grave is guaranteed with no requirement that you actually need to work. Once again we have to wonder where the money will come from to fulfill this guarantee . If no one has to work, there won't be anyone to collect taxes from. It's all a little confusing—harebrained or half-baked would be a better description.

# CHAPTER 5

# THE HELL OF SOCIALISM

On the surface it's hard to criticize the concept of socialism. That's the reason it's had such a grip on the minds of men for centuries. Who can be against a political system predicated on the equal sharing of the fruits of the community's labor? Equality for all. No one profiting off the labor of another human being. All offering each other a helping hand. A community working harmoniously together. Certainly, these are the precepts that should guide our lives. Sounds wonderful. . . .and then we awaken from the dream . . . and face reality.

There may be one person in a thousand that could actually function in the world described above. These people normally enter a religious life, such as Mother Theresa. The discussion of the merits of capitalism and socialism has become an adversarial contest between head-in-the-clouds intellectuals and the down-to-earth pragmatists. The simple fact about man's nature is this—we each strive to better our own personal and family's condition above that of the community. We may make efforts as required as a result of human compassion but in the final analysis we are looking after our own well-being.

These statements on the nature of man are not original to this time. Adam Smith, an eighteenth century economist, stated the underlying fallacy of socialism in his landmark book "The Wealth of Nations" :

> *"It is not from the benevolence of the butcher, the brewer, or the baker, that we expect our dinner, but from their regard to their own self-interest."*

History has proven that socialism as a political system has never worked, in its many variations—from the socialistic underpinning of fascism epitomized by the Nazis, to the totalitarian regimes of Mao Tse Tung, Joseph Stalin, Pol Pot, Fidel Castro, Hugo Chavez and the dynasty of North Korea. We're told, time and time again, that these socialistic examples are outliers, and that true socialism has never been given a chance. Many point to the Scandinavian countries but those countries' leaders and their citizens emphatically state they embrace capitalism. In the few instances where the some of the Scandinavian political parties tried to introduce stronger socialistic policies they were voted out of office. People want freedom and true socialism will not allow it.

Socialistic governments can only rule by force. Monetary incentives are anathema to socialistic governments but most find them necessary in some form for the society's survival. Time and time again it has been found that free-market capitalistic societies provide a higher standard of living for the majority of its citizens.

However, there are tradeoffs between socialistic and capitalistic societies. In fact, a form of economic Darwinism comes into play in a capitalistic society. It's an old adage that cream rises to the top. A capitalistic society rewards those who take chances, work hard or come up with new ideas. A familiar quote by Winston Churchill puts it all very succinctly:

> "The inherent vice of capitalism is the unequal sharing of blessings. Socialism is a philosophy of failure, the creed of ignorance and the gospel of envy; its inherent virtue is the equal sharing of miseries."

In essence, under a capitalistic economy there are winners and losers—although even the "losers" are better off economically than under socialism. In a socialistic economy there are only losers— except for those at the very top of the ladder, the governing few enforcing the laws of the society. The important point to be made here is that under a capitalistic society there are many winners so the entire society is lifted up- comparable to a rising tide lifting all boats.

The key requirement in a vibrant capitalistic economy is "opportunity". Substantially all of its citizens must have the opportunity to better themselves economically. Without true opportunity for its citizens a capitalistic society will fail—either through evolution or revolution.

Another quote, from Daniel Pipes, seems appropriate in closing this chapter:

> *"If bad ideas bring horror, their antidote lies in conservative, modest, tried-and-tested ideas that respect tradition and human nature, not in revolutionary lurches and grandiose experiments, but in incremental improvements in customary practices. At a moment when many Democrats are ignoring the lessons of Venezuela and swooning over socialism, it's back to the barricades in the war of ideas."*

In 1950 Venezuela enjoyed the fourth highest per capita income in the world due to its possession of the largest oil reserves on earth. Today, due to the imposition of a despotic socialist agenda its people are starving, disease and criminality are rampant, and there is mass migration out of the country.

Finally, Governor Nikki Haley recently gave an excellent speech comparing capitalism and socialism to the Hudson Institute. It is attached as Appendix I, and I encourage you to read it.

# CHAPTER 6

# THE DEVIL'S TWIN BROTHER: THE PARIS CLIMATE ACCORD

The Paris Climate Accord is an agreement signed by 200 nations in 2015 that ostensibly would require all of the signatories to take actions to prevent climate change. To facilitate the efforts of other countries the United States would be required to subsidize the efforts of other nations.

The Accord is attached as Appendix II following the Relevant Quotes section... The reader will soon discover after attempting to decipher its enigmatic and verbose text that it actually doesn't require any country, including two of the highest polluters in the world (China and India) to do anything other than mouth meaningless goals. Each country in the world is asked to submit its plans for CO2 reduction, but binds them to nothing. The Obama administration, on the other hand, promised to take strong actions that would undermine the U. S. economy and to contribute billions of dollars to help other countries, including China and India, to reach some ephemeral goal—the very definition of throwing money down a rathole.

Realizing the ludicrous provisions of this agreement President Trump withdrew the United States from the agreement when he came into office in 2017 with the following statement:

35

*Statement by President Trump on the Paris Climate Accord (abbreviated)*

"In order to fulfill my solemn duty to protect America and its citizens, the United States will withdraw from the Paris Climate Accord — but begin negotiations to reenter either the Paris Accord or a really entirely new transaction on terms that are fair to the United States, its businesses, its workers, its people, its taxpayers. So we're getting out. But we will start to negotiate, and we will see if we can make a deal that's fair. And if we can, that's great. And if we can't, that's fine.

"As President, I can put no other consideration before the wellbeing of American citizens. The Paris Climate Accord is simply the latest example of Washington entering into an agreement that disadvantages the United States to the exclusive benefit of other countries, leaving American workers — who I love — and taxpayers to absorb the cost in terms of lost jobs, lower wages, shuttered factories, and vastly diminished economic production.

"Thus, as of today, the United States will cease all implementation of the non-binding Paris Accord and the draconian financial and economic burdens the agreement imposes on our country. This includes ending the implementation of the nationally determined contribution and, very importantly, the Green Climate Fund which is costing the United States a vast fortune.

"Compliance with the terms of the Paris Accord and the onerous energy restrictions it has placed on the United States could cost America as much as 2.7 million lost jobs by 2025 according to the National Economic Research Associates. This includes 440,000 fewer manufacturing jobs — not what we need — believe me, this is not what we need — including automobile jobs, and the further decimation of vital American industries on which countless communities rely. They rely for so much, and we would be giving them so little.

"According to this same study, by 2040, compliance with the commitments put into place by the previous administration would cut production for the following sectors: paper down 12 percent; cement down 23 percent; iron and steel down 38 percent; coal — and I happen to love the coal miners — down 86 percent; natural gas down 31 percent. The cost to the economy at this time would be close to $3 trillion in lost GDP and 6.5 million industrial jobs, while households would have $7,000 less income and, in many cases, much worse than that.

"Not only does this deal subject our citizens to harsh economic restrictions, it fails to live up to our environmental ideals. As someone who cares deeply about the environment, which I do, I cannot in good conscience support a deal that punishes the United States — which is what it does -- the world's leader in environmental protection, while imposing no meaningful obligations on the world's leading polluters.

"For example, under the agreement, China will be able to increase these emissions by a staggering number of years — 13. They can do whatever they want for 13 years. Not us. India makes its participation contingent on receiving billions and billions and billions of dollars in foreign aid from developed countries. There are many other examples. But the bottom line is that the Paris Accord is very unfair, at the highest level, to the United States.

"Further, while the current agreement effectively blocks the development of clean coal in America — which it does, and the mines are starting to open up. We're having a big opening in two weeks. Pennsylvania, Ohio, West Virginia, so many places. A big opening of a brand-new mine. It's unheard of. For many, many years, that hasn't happened. They asked me if I'd go. I'm going to try.

"China will be allowed to build hundreds of additional coal plants. So we can't build the plants, but they can, according to this agreement. India will be allowed to double its coal production by 2020. Think of it: India can double their coal production. We're supposed to get rid of ours. Even Europe is allowed to continue construction of coal plants.

"In short, the agreement doesn't eliminate coal jobs, it just transfers those jobs out of America and the United States, and ships them to foreign countries.

"This agreement is less about the climate and more about other countries gaining a financial advantage over the United States. The rest of the world applauded when we signed the Paris Agreement — they went wild; they were so happy — for the simple reason that it put our country, the United States of America, which we all love, at a very, very big economic disadvantage. A cynic would say the obvious reason for economic competitors and their wish to see us remain in the agreement is so that we continue to suffer this self-inflicted major economic wound. We would find it very hard to compete with other countries from other parts of the world.

"We have among the most abundant energy reserves on the planet, sufficient to lift millions of America's poorest workers out of poverty. Yet, under this agreement, we are effectively putting these reserves under lock and key, taking away the great wealth of our nation — it's great wealth, it's phenomenal wealth; not so long ago, we had no idea we had such wealth — and leaving millions and millions of families trapped in poverty and joblessness.

"The agreement is a massive redistribution of United States wealth to other countries. At 1 percent growth, renewable sources of energy can meet some of our domestic demand, but at 3 or 4 percent growth, which I expect, we need all forms of available American energy, or our country — will be at grave risk of brownouts and blackouts, our businesses will come to a halt in many cases, and the American family will suffer the consequences in the form of lost jobs and a very diminished quality of life.

"Even if the Paris Agreement were implemented in full, with total compliance from all nations, it is estimated it would only produce a two-tenths of one degree — think of that; this much — Celsius reduction in global temperature by the year 2100. Tiny, tiny amount. In fact, 14 days of carbon emissions from China alone would wipe out the gains from America — and this is an incredible statistic — would totally wipe out the gains from America's expected reductions in the year 2030, after we have had to spend billions and billions of dollars, lost jobs, closed factories, and suffered much higher energy costs for our businesses and for our homes.

"As the Wall Street Journal wrote this morning: 'The reality is that withdrawing is in America's economic interest and won't matter much to the climate.' The United States, under the Trump administration, will continue to be the cleanest and most environmentally friendly country on Earth. We'll be the cleanest. We're going to have the cleanest air. We're going to have the cleanest water. We will be environmentally friendly, but we're not going to put our businesses out of work and we're not going to lose our jobs. We're going to grow; we're going to grow rapidly.

"I'm willing to immediately work with Democratic leaders to either negotiate our way back into Paris, under the terms that are fair to the United States and its workers, or to negotiate a new deal that protects our country and its taxpayers.

"So if the obstructionists want to get together with me, let's make them non-obstructionists. We will all sit down, and we will get back into the deal. And we'll make it good, and we won't be closing up our factories, and we won't be losing our jobs. And we'll sit down with the Democrats and all of the people that represent either the Paris Accord or something that we can do that's much better than the Paris Accord. And I think the people of

41

our country will be thrilled, and I think then the people of the world will be thrilled. But until we do that, we're out of the agreement.

"I will work to ensure that America remains the world's leader on environmental issues, but under a framework that is fair and where the burdens and responsibilities are equally shared among the many nations all around the world.

'No responsible leader can put the workers — and the people — of their country at this debilitating and tremendous disadvantage. The fact that the Paris deal hamstrings the United States, while empowering some of the world's top polluting countries, should dispel any doubt as to the real reason why foreign lobbyists wish to keep our magnificent country tied up and bound down by this agreement: It's to give their country an economic edge over the United States. That's not going to happen while I'm President. I'm sorry.

"My job as President is to do everything within my power to give America a level playing field and to create the economic, regulatory and tax structures that make America the most prosperous and productive country on Earth, and with the highest standard of living and the highest standard of environmental protection.

"The Paris Agreement handicaps the United States economy in order to win praise from the very foreign capitals and global activists that have long sought to gain wealth at our country's expense. They don't put America first. I do, and I always will.

"The same nations asking us to stay in the agreement are the countries that have collectively cost America trillions of dollars through tough trade practices and, in many cases, lax contributions to our critical military alliance. You see what's happening. It's pretty obvious to those that want to keep an open mind.

"At what point does America get demeaned? At what point do they start laughing at us as a country? We want fair treatment for its citizens, and we want fair treatment for our taxpayers. We don't want other leaders and other countries laughing at us anymore. And they won't be. They won't be.

"I was elected to represent the citizens of Pittsburgh, not Paris. I promised I would exit or renegotiate any deal which fails to serve America's interests. Many trade deals will soon be under renegotiation. Very rarely do we have a deal that works for this country, but they'll soon be under renegotiation. The process has begun from day one. But now we're down to business.

"Beyond the severe energy restrictions inflicted by the Paris Accord, it includes yet another scheme to redistribute wealth out of the United States through the so-called Green Climate Fund — nice name — which calls for developed countries to send $100 billion to developing countries all on top of America's existing and massive foreign aid payments. So we're going to be paying billions and billions and billions of dollars, and we're already way ahead of anybody else. Many of the other countries haven't spent anything, and many of them will never pay one dime.

"The Green Fund would likely obligate the United States to commit potentially tens of billions of dollars of which the United States has already handed over $1 billion — nobody else is even close; most of them haven't even paid anything — including funds raided out of America's budget for the war against terrorism. That's where they came. Believe me, they didn't come from me. They came just before I came into office. Not good. And not good the way they took the money.

"In 2015, the United Nation's departing top climate officials reportedly described the $100 billion per year as "peanuts," and stated that "the $100 billion is the tail that wags the dog." In 2015, the Green Climate Fund's executive director reportedly stated that estimated funding needed would increase to $450 billion per year after

2020. And nobody even knows where the money is going to. Nobody has been able to say, where is it going to?

"Of course, the world's top polluters have no affirmative obligations under the Green Fund, which we terminated. America is $20 trillion in debt. Cash-strapped cities cannot hire enough police officers or fix vital infrastructure. Millions of our citizens are out of work. And yet, under the Paris Accord, billions of dollars that ought to be invested right here in America will be sent to the very countries that have taken our factories and our jobs away from us. So think of that.

"There are serious legal and constitutional issues as well. Foreign leaders in Europe, Asia, and across the world should not have more to say with respect to the U.S. economy than our own citizens and their elected representatives. Thus, our withdrawal from the agreement represents a reassertion of America's sovereignty. Our Constitution is unique among all the nations of the world, and it is my highest obligation and greatest honor to protect it. And I will.

"Staying in the agreement could also pose serious obstacles for the United States as we begin the process of unlocking the restrictions on America's abundant energy reserves, which we have started very strongly. It would once have been unthinkable that an international agreement could prevent the United States from conducting its own domestic economic affairs, but this is the new reality we face if we do not leave the agreement or if we do not negotiate a far better deal.

"The risks grow as historically these agreements only tend to become more and more ambitious over time. In other words, the Paris framework is a starting point — as bad as it is — not an end point. And exiting the agreement protects the United States from future intrusions on the United States' sovereignty and massive future legal liability. Believe me, we have massive legal liability if we stay in.

"As President, I have one obligation, and that obligation is to the American people. The Paris Accord would undermine our economy, hamstring our workers, weaken our sovereignty, impose unacceptable legal risks, and put us at a permanent disadvantage to the other countries of the world. It is time to exit the Paris Accord — and time to pursue a new deal that protects the environment, our companies, our citizens, and our country.

"It is time to put Youngstown, Ohio, Detroit, Michigan, and Pittsburgh, Pennsylvania — along with many, many other locations within our great country — before Paris, France. It is time to make America great again. "

President Trump's statement says it all, far better than I could.

Unfortunately, elections have consequences, and upon entering office in 2021, President Biden, by executive order, had the United States rejoin this disastrous one-sided agreement. He also appointed John Kerry, who negotiated the abysmal Iran Nuclear Agreement as his "Climate Czar", and instructed him to go to China and plead with them to reduce their emissions. Fat chance of that happening with their plans for the construction of numerous coal plants, and their ability to increase $CO_2$ emissions until 2030 under the Paris Accord since they're considered an "underdeveloped nation." The irony of the situation must have the Chinese laughing at our stupidity

# CHAPTER 7

## THE STRUGGLE AHEAD

It would be the easiest course to throw up our hands and say there's little we can do to stop the implementation of the programs of the progressive left given that they currently control both houses of Congress and the presidency. However, even faced with what seem like insurmountable obstacles there are actions that can be taken.

We must never give up! Call and write your congressional representatives and senators, when appropriate. Politically and financially support advocates for a free market economy whenever possible. Talk to your friends and neighbors. Become involved in your local politics, including the school boards. Much of the propaganda about global warming, the slanted history of the country's founding and the evils of a capitalistic form of government, now being taught at the college level, have crept into our local grade and high schools. Find out what your children are being taught and confront the school authorities, when necessary. Many parents have done that, and have been successful in removing objectionable material. If necessary, transfer your children from the public school system and put them in a private school or home school them. The important point to be made here is to take action. Make your voice heard. You must speak up.

We are engaged in a struggle to the death with those who want to force a socialistic dictatorship upon the United States.

The sooner we wake up to that fact, the better chance we will have of stopping the progressives plan to impose socialism on the United States.

It seems appropriate here to end with an excerpt from a poem which was quoted by President Reagan, some years ago, after a political set-back:

*I'm a little wounded, but I am not slain*

*I will lay me down to bleed a while*

*Then I'll rise and fight again.*

*-- John Dryden*

# RELEVANT QUOTES

"The whole aim of practical politics is to keep the populace alarmed (and clamorous to be led to safety} by menacing it with an endless series of hobgoblins, all of them imaginary"—H. L. Mencken. Man-made climate change is the latest hobgoblin foisted on the American public.

"Global warming is largely a natural phenomenon. The world is wasting stupendous amounts of money on trying to fix something that can't be fixed."—Dr. David Bellamy, Botanist and Environmentalist

" . . . the innate behavior of the climate system imposes limits on the ability to predict its evolution."—From the latest report of the Inter-governmental Panel on Climate Change, published in March 2014. This is an astounding statement by the IPCC, given its years of wringing hands and caterwauling. about the impending doom of the planet due to man-made global warming, contrary to the evidence showing no significant global warming.

"When we make (mathematical) models involving human beings we are trying to force the ugly stepsister's foot into Cinderella's pretty glass slipper. It doesn't fit without cutting out some of the essential parts."—Emanuel Derman, from his book, *Models Behaving Badly*. This analogy also applies to the climate change models in use today, none of which has been proven accurate.

"I used to agree with these dramatic warnings of climate disaster . . . However, a few years ago I decided to look more closely at the science and it astonished me. In fact, there is no evidence of humans being the cause. There is, however, overwhelming evidence of natural causes such as changes in the output of the sun."—Dr. Ian D. Clark, Paleo-climatologist

"The causes of these global changes is fundamentally due to the Sun and its effect on the Earth as it moves about in its orbit—not from man-made activities."—Dr. William W. Vaughn, Award Winning NASA Atmospheric Scientist

"Satellites have recorded a roughly 14% increase in greenery on the planet over the past 30 years, in all types of ecosystems, partly as a result of man-made CO2 emissions, which enable plants to grow faster and use less water. . . Almost every global environmental scare of the past half-century proved exaggerated, including the population "bomb", pesticides, acid rain, the ozone hole, falling sperm counts, genetically engineered crops and killer bees. In every case, institutional scientists gained a lot of funding from the scare and then quietly converged on the view that the problem was much more moderate than the extreme voices had argued. Global warming is no different."—Matt Ridley

"Scratch an intellectual, and you will find a would-be aristocrat who loathes the sight, the sound and the smell of common folk."—Eric Hoffer, from his book *The True Believer*

"If we extend unlimited tolerance even to those who are intolerant, if we are not prepared to defend a tolerant society against the onslaught of the intolerant, then the tolerant will be destroyed and tolerance with them."—Karl Popper, explaining "the paradox of tolerance"

"How can you say to the hungry of this earth—how can you say to those who don't enjoy the luxury that we all do and that the developed world in general does, how can you tell those folks, 'Sorry about your luck.' You know this is an indulgence of the rich and it is not just scientifically indefensible, it is morally indefensible."—From a speech by Purdue University President Mitch Daniels, Jr., attacking the anti-GMO movement in its attempts through junk science and false claims to stifle new technologies which would allow food abundance for all. The same statement is applicable to those who would deny oil, gas and coal to provide economic power to these underdeveloped economies, where a significant part of their populace is without electricity.

"Banning DDT killed more people than Hitler."—Author Michael Crichton, alluding to the fact that the DDT ban allowed mosquitoes to infect the malaria virus into millions of African children. Another calamitous example of the rule of unintended consequences of precipitous actions taken to solve current problems.

"I firmly believed that the ends justified the means. Our great goal was the universal triumph of Communism, and for the sake of the goal everything was permissible—to lie, to steal, to destroy hundreds of thousands and even millions of people, all those that were hindering our work, or could hinder it, everyone who stood in our way. And to hesitate and doubt about all this was to give in to 'intellectual squeamishness' and 'stupid liberalism'."—Said by a Russian Communist, following the 1917 revolution. It appears that this is now the attitude of the man-made climate fanatics pushing for the Green New Deal and the Paris Climate Accord.

"Some might say the USSR and Venezuela never implemented true socialism. Even so, that merely confirms another argument against socialism: it has to rely on imperfect, self-interested people to staff its bureaucracies, plan the economy, and oversee an excessively centralized administration. Dreams of liberation and progress quickly deteriorate into tyranny and dictatorship. Once fallible men get a taste of power, they are reluctant to relinquish it."—Jeff Cimino

"Since heat-related deaths are generally much fewer than cold-related deaths, the overall effect of global warming on health can be expected to be a beneficial one."—From a 2004 study by William Keating and Gavin Donaldson

"The first lesson of economics is scarcity: There is never enough of anything to satisfy all those who want it. The first lesson of politics is to ignore the first lesson of economics."—Thomas Sowell.

"A civilization that feels guilty for everything it is and does will lack the energy and conviction to defend itself."—French philosopher Jean Francois Revel. Quoted by United Nations Ambassador Jeanne Kirkpatrick, a lifelong Democrat, as she electrified the 1984 Republican National Convention, when she described the Democratic National Convention, held some weeks earlier, as a meeting of the "Blame America First" crowd. This "crowd" has now returned with a vengeance in the 21st century, blaming America for much of the global warming they say is occurring.

"(Socialism that is democratic) has never existed, and in its absence socialism without democracy becomes the shimmering goal (of many liberals), because socialism cannot be given up. Therein lies the final irony. Just as democratic socialism is an illusion, so too is socialism itself the pipe dream that never dies. It promises harmony and abundance; it has always instead produced strife and penury. Little wonder, then, that free people have never chosen it."—Joshua Muravchik, in a commentary in *The Weekly Standard.*

"Soviet life was a pail of milk of human kindness, with a dead rat at the bottom."—Russian writer Vladimir Nabokov

"There is no Plan B because there is no Planet B."—French President Emmanuel Macron lecturing President Donald Trump, after Trump had announced that the United States was withdrawing from the Paris climate agreement. Within months Macron had backtracked from his plan to impose new carbon taxes on the French people as part of his fulfillment of the climate agreement's requirements, in the face of massive riots throughout France. Apparently there will have to be a Plan B.

"Socialist economic proposals are recipes for economic stagnation.  If the state owns corporations, there is no competition, only rivalries among people with political power."—Stephen Miller

"My reading of history convinces me that most bad government results from too much government."—President Thomas Jefferson

"I regard consensus science as an extremely pernicious development that ought to be stopped cold in its tracks.  Historically, the claim of consensus has been the first refuge of scoundrels; it is a way to avoid debate by claiming that the matter is already settled.  Whenever you hear the consensus of scientists agrees on something or other, reach for your wallet, because you're being had. . . . There is no such thing as consensus science.  If it's consensus, it isn't science. . . .Nobody believes a weather prediction twelve hours ahead.  Now we're asked to believe a prediction that goes out 100 years into the future?  And make financial investments based on that prediction?  Has everybody lost their minds?"—Author Michael Crichton

"All great movements start as a cause, evolve into a business and end up a racket."—Eric Hoffer, author of *The True Believer*. *Man-made climate change is the latest example of this truism.*

"Underlying most arguments against the free market is a lack of belief in freedom itself."—Milton Friedman

"Here's the truth, brothers and sisters, there's plenty of money in the world. There's plenty of money in this city. It's just in the wrong hands."—New York City Mayor Bill de Blasio, making a statement typical of a socialist redistributionist who has zero understanding of economics. How many of those who earned the money are going to stick around waiting for him to confiscate it?

"All active mass movements strive to interpose a fact-proof screen between the faithful and the realities of the world. They do this by claiming that the ultimate and absolute truth is already embodied in their doctrine and that there is no truth nor certitude outside it. To rely on the evidence of the senses and reason is heresy and treason."—Eric Hoffer "*The True Believer*"

"You're going to find signs on manufacturing doors, if this bill passes, that say, 'Moved—gone to China'."—Senator Charles Grassley, commenting on the cap-and-trade bill, coming before the Senate, designed to curb greenhouse-gas emissions, by imposing significant costs on manufacturing industries using oil and coal.

"We now know that the extra carbon dioxide and global warmth, no matter what their cause, are resulting in a gradual greening of the Earth. There is some evidence that there has been a slight poleward shift in the habitats of some warm weather species, from the tropics where there is a great diversity of life, to higher latitudes where many of these forms of life could not otherwise survive. Global warming has made weather less severe, and cold weather is known to cause more deaths than hot weather. So why is global warming necessarily a bad thing?"—Roy W. Spencer, from his book *Climate Confusion*.

"If we were directed from Washington when to sow and when to reap, we would soon want bread."—Thomas Jefferson

"The link between the burning of fossil fuels and global warming is a myth. It is time the world's leaders, their scientific advisers and many environmental pressure groups woke up to the fact"

* * *

"...carbon dioxide is *not* the dreaded killer greenhouse gas...It is, in fact, the most important airborne fertilizer in the world... (vital for photosynthesis).

* * *

"The real truth is that the main greenhouse gas—the one that has the most direct effect on land temperatures—is water vapour, 99% of which is entirely natural."—Excerpts from a full page article titled "Global Warming? What a Load of Poppycock!" in the London newspaper, *The Daily Mail*, by Professor David Bellamy, an ardent environmentalist.

"Workers of the world, forgive me."—Graffiti written on the bust of Karl Marx, founder of Communism, in Bucharest, Romania, in a play on the words of the Communist Manifesto, "Workers of the World, Unite!"

"We're going to need oil and gas and coal for a long time if America wants to keep the lights on. What I see are people who want affordable energy. They want strong environmental standards—they want a lot of things—but first and foremost they want affordable energy. And if you want affordable energy, you want oil, gas and coal."—John Watson, CEO, Chevron Oil.

"Actually what makes skeptics (of global warming) skeptical is the accumulating evidence that theories predicting catastrophe from man-made climate change are impervious to evidence."—George Will, 2009, commenting on the trend reported in the *New York Times* that, "...global temperatures have been relatively constant for a decade and may even drop in the next few years."

"The sudden abundance of low-cost natural gas is a gift."—Tim Wirth, president of the United Nations Foundation. With the introduction of a new drilling technique (hydraulic fracturing) developed in the late 1990s, immense deposits of shale gas became recoverable in the United States. The quantity of recoverable natural gas in the United States and Canada is now estimated to last 100 years.

"Atlas Shrugged."—Title of Ayn Rand's bestseller that praised unrepentant capitalism, published in 1957. The novel's title refers to the action of the mythological Atlas who carried the world on his shoulders until the weight became too much. The book's underlying thesis was that self-interest is paramount to all other interests and that when a society restricts and taxes its most productive and creative citizens too much in the interest of wealth redistribution, these citizens will simply "throw in the towel", walk away and let the society perish. The book sold 400,000 copies in 2009, double the total of previous sales in any of the previous fifty-two years since its publication.

"When you see that in order to produce you must obtain permission from men who produce nothing . . . and your laws don't protect you against them, but protect them against you . . . you may know that your society is doomed."—From *Atlas Shrugged* by Ayn Rand.

"Communism is the time that countries waste between capitalism and capitalism."—Cuban Carlos Alberto Montaner, commenting on the 52nd year that the Cuban government has wasted waiting for prosperity.

"The capitalistic economy goes out of its way to put ourselves first. (It derives from an egotism) rooted in the 'old brain' which was bequeathed to us by the reptiles that struggled out of the primal slime 500 million years ago."—Karen Armstrong, author of *Twelve Steps to a Compassionate Life.*

". . . those nasty old capitalists, with their vigor, risk-taking, animal spirits and reptilian brains, have created so much wealth for so many societies over so many centuries—and have raised the standard of living for so many people who would otherwise live in grinding poverty—that their efforts, easily considered merely selfish, begin to look downright compassionate."—Eric Felton, in his critical review of the book cited in the previous quote above, in *The Wall Street Journal.*

"Are we never to learn that socialism has its roots in envy and in nothing else."—Norman Douglas

"The most astounding fact about reformers, driven by the purest of motives and most spotless goodwill, is that it does not dawn on them that their programs can make things worse."—Leo Rosten

". . . China and India, . . . have told Obama officials they have no intention of signing on to the rich world's growth-killing obsessions."—*Wall Street Journal* editorial response to the projected U.S. cap-and-tax bill on carbon emissions expected to drive up the cost of U.S. manufacturers and ultimately to result in the movement of the manufacturers overseas. A group of U. S. Senators from states with significant manufacturing employment indicated they wanted a tariff on imports from countries without carbon emissions restrictions, as the price for their support of the bill—an action that would result in a trade war.

"Congress . . . will spurn calls to send billions in 'climate reparations' to China and other countries. Representatives of those nations, when they did not have their hands out in Copenhagen, grasping for America's wealth, clapped their hands for Hugo Chavez and other kleptocrats who denounced capitalism while clamoring for its fruits."—George Will, in a commentary on the Copenhagen Climate Conference.

"Global warming is necessary to prevent a new Ice Age."—Conclusion of an article in *Science* magazine (January 2010).

"We can't wait to hear Mr. Obama (and now Mr. Biden) tell Americans that he wants them to pay higher taxes so the U. S. can pay China to become more energy efficient and thus more economically competitive."—From a *Wall Street Journal* editorial commenting on the announcement of an initiative by the Obama administration at a climate conference in Denmark to help raise an annual $100 billion fund to assist developing nations, including China, to become more energy efficient and thus help reduce "man-made" global warming, an effort climatologists agree would have little or no impact on the earth's climate. In early 2021 President Biden caused the United States to reenter the Paris Climate Accord reactivating Obama's earlier promise.

"The frog is coming awake at just the last moment. He is jumping out of the water."—Peggy Noonan, comparing the American people, on the issue of excessive government spending, to a frog in a pot of water—it was predicted that the rising heat would lull the frog, and when the water came full boil, it wouldn't be able to jump out. This "jumping frog" analogy could just as well be applicable to the Green New Deal.

"Occupy Everything, Death to Capitalism." —Large black banner carried by protesters in Oakland, California. My question is "Why are they here?" They should all be applying for visas to go to North Korea, the anti-capitalist paradise of their dreams.

"Democrats believe in the welfare state before they believe in capitalism. The assumption is that there is some kind of perpetual engine of economic prosperity in America that is just going to continue. And therefore they are able to take from those who create and give it to those who don't."—Republican Eric Cantor, House Majority Leader

"So that the record of history is absolutely crystal clear. That there is no alternative way, so far discovered, of improving the lot of the ordinary people that can hold a candle to the productive activities that are unleashed by a free enterprise system."—Milton Friedman

"Somehow we have got to figure out how to boost the price of gasoline to the levels in Europe."— Steven Chu, U. S. Secretary of Energy, under the Obama administration.

"Everyone must be the same and have the same. Social justice means we deny ourselves many things so that others may have to do without them as well."—Sigmund Freud, commenting on class envy.

"I have been a member of the European Parliament for twelve years. I am living in your future or at least the future your present leaders seem intent on taking you. Believe me, my friends, you are not going to enjoy it. . . . We are at the end of the road you have just set out along. . . . We're screeching towards the cliff. You know what? We look up and what do we see in our rear view mirror? We see you trying to overtake us, accelerating frantically in the direction we have been going. My friends, there is still time to turn aside."—Daniel Hannon, one of Great Britain's representatives to the European Parliament, and one of the few dissenting members to its current policies, in a 2012 speech to CPAC (Conservative Policy Action Council) in Washington, D. C.

"Socialists can provide you shelter, fill your belly with bacon and beans, treat you when you're ill— all the things guaranteed to a prisoner or a slave."—Ronald Reagan

"  .  .  .  the greatest and most successful pseudoscientific fraud I have seen in my long life as a physicist."—Harold Lewis, University of California Emeritus Professor of Physics, castigating the science behind manmade global warming. Ivar Giaever, 1976 Nobel Laureate, supporting Lewis, stated that a .8 degree change, on the Kelvin scale from -288.0 to -288.8 in 150 years, meant to him that the earth's temperature was amazingly stable.

"Environmental regulations are seen to be the number one risk to reliability over the next one to five years."—From the report of the North American Electric Reliability Corporation (NERC), an independent advisory body, appointed by Congress, to monitor the reliability of the nation's electrical grid. Stung by the criticism, Obama administration EPA officials initiated an audit of the NERC, in a "kill the messenger" reaction.

"As Americans we must always remember we all have a common enemy, an enemy who is dangerous, powerful and relentless. I refer, of course, to the federal government."—Dave Barry. Marquette University demanded that one of its graduate students remove this quotation, which had been posted on the student's door.

"Can you imagine 400 million people who do not have a light bulb in their home? You cannot, in a democracy, ignore some of these realities and as it happens with the resources of coal that India has, we really don't have any choice but to use coal."—Rajandra Pachauri, Indian academic, chairman of the United Nations Intergovernmental Panel On Climate Change, facing the reality that coal will not be displaced as one of the primary sources of electrical generation anytime in the foreseeable future.

"Socialism is like a dream. Sooner or later you wake up to reality."—Winston Churchill

"If you think health care is expensive now, wait until you see what it costs when it's free!"—P. J. O'Rourke

"You are never dedicated to something you have complete confidence in. (No one is fanatically shouting that the sun is going to rise tomorrow. They know it's going to rise tomorrow.) When people are fanatically devoted to political or religious faiths or any other kinds of dogma or goals, it's always because these dogmas or goals are in doubt."—Robert M. Pirsig

"I wish that during the years that I was in public office, I had had this firsthand experience. We intuitively know that to create job opportunities, we need entrepreneurs who will risk their capital against an expected payoff. Too often, however, public policy does not consider whether we are choking off those opportunities."—Former Democratic presidential candidate George McGovern, who had recently lost his entire investment in a Connecticut motor inn, which he attributed to burdensome governmental regulations.

"One has to belong to the intelligentsia to believe things like that: no ordinary man could be such a fool."—George Orwell, commenting on some of the theories of contemporary scientists. Quotes such as this are timeless.

"The pause in the rise of the global average temperature may have already lasted 17 years, depending on which data set you look at."—Rajendra Pachauri, chairman of the Intergovernmental Panel on Climate Change. Based on hindsight, some climate change models have overestimated warming by 100% over the past 20 years.

"It comes from outer space, lands in the woods, and it's the size of an orange or cantaloupe. By the end of the movie, enveloping diners and houses. As the blob rolled along eating folks, it got bigger and bigger. It never got smaller. And that's how I kind of see the federal government. Instead of eating people and diners, it's eating our liberty. I'm not against government. I'm against this ever-expanding government that doesn't know its limits."—South Carolina attorney general Alan Wilson, using the symbolism of the movie *The Blob*, to describe the ever encroaching reach of the federal government.

"I think we have to stop considering *Climate Research* as a legitimate peer-review journal ... Perhaps we should encourage our colleagues in the climate research community to no longer submit to, or cite papers in, this journal."—Climate researcher Professor Michael Mann to a colleague in a leaked e-mail, commenting on the publication of an article questioning global warming theories in the cited journal. Professor Mann was the creator of the infamous "hockey stick" graph showing a sudden upward turn in world average temperature since the Industrial Revolution—a depiction that was later shown to be based on erroneous statistics. This bogus graph was used prominently in former Vice President Gore's award winning documentary "An Inconvenient Truth."

"The (environmental) prophet is not a great soul who admonishes us but a petty fellow who wishes us many misfortunes if we have the gall not to listen to him. Catastrophe is not something that haunts him but his source of joy."—Pascal Bruckner, from his book *The Fanaticism of the Apocalypse*.

"All the foolishness of Bolshevism, Maoism, and Trotskyism are somehow reformulated exponentially in the name of saving the planet."—Pascal Bruckner, from his book *The Fanaticism of the Apocalypse*, descrying the environmentalists of today.

"Just coal by another name."—Environmental activist Mike Tidwell, describing natural gas, ignoring the fact that the vast quantities of natural gas released by the "fracking" method in the United States has resulted in an 11% decrease in carbon dioxide emissions between 2005 and 2011 due to the replacement of coal in power plants. Beside ignoring this fact Tidewell and other activists called for a substantial and costly increase in renewable sources such as solar and wind. Germany, which poured billions into renewable sources has seen its electrical costs soar, and has now turned to coal as its primary source of electricity production.

"A society that puts equality—in the sense of equality of outcome—ahead of freedom will end up with neither equality nor freedom. The use of force to achieve equality will destroy freedom, and the force, introduced for good purposes, will end up in the hands of people who use it to promote their own interests. . . Freedom means diversity but also mobility. It preserves the opportunity for today's disadvantaged to become tomorrow's privileged and, in the process, enables almost everyone, from top to bottom, to enjoy a fuller and richer life."—Milton and Rose Friedman from *Free To Choose*.

"Instead of fostering a system that enables people to help themselves, America is now saddled with a system that destroys value, raises costs, hinders innovation and relegates millions of citizens to a life of poverty, dependency and hopelessness. This is what happens when elected officials believe that people's lives are better run by politicians and regulators than by the people themselves. Those in power fail to see that more government means less liberty, and liberty is the essence of what it means to be American. Love of liberty is the American ideal."—Charles O. Koch, chairman and CEO of Koch Industries, in a criticism written during the Obama administration—and now applicable to the Biden administration

"Observe which side resorts to the most vociferous name-calling and you are likely to have identified the side with the weaker argument and they know it."—Charles R. Anderson, Research Physicist

"You had designers who were constrained and occupied with only one goal, and that was weight and miles per gallon."—Sam Kazman, general counsel of the Competitive Enterprise Council, strongly criticizing the CAFE standards that require higher "miles per gallon" on manufactured cars. According to a 2007 Insurance Institute for Highway Safety 250-500 deaths are attributable to downsized cars attempting to meet more stringent CAFE standards.

"It is not from the benevolence of the butcher, the brewer, or the baker, that we expect our dinner, but from their regard to their own self-interest."—Adam Smith, eighteenth century economist, and author of *The Wealth of Nations*, who theorized that an "invisible hand" underpinned the actions of participants in a market economy.

"(Environmentalism has) become a religion, and religions don't worry much about facts."—James Lovelock, environmental scientist

"In one of the most expensive ironies of history, the expenditure of more than $50 billion on research into global warming has failed to demonstrate any human-caused climate trend, let alone a dangerous one."—Robert Carter, Paleoclimate Scientist"


"Eat less!"—Dame Vivienne Westwood, millionaire British fashion designer, and advocate against genetically modified food, when asked what poor people should do if they couldn't afford to buy organic food. Her response is indicative of the attitude of rich, effete snobs who look down on people who can't afford the luxury of subscribing to their elitist views.


When Milton (Friedman) was starting out, people really believed a state run economy was the most efficient way of promoting growth. Today nobody believes that, except maybe in North Korea. You go to China, India, Brazil, Argentina, Mexico, even Western Europe. Most of the economists under 50 have a free market orientation. Now, there are differences of emphasis and opinion among them. But they're oriented toward the markets. That's a very, very important victory. Will this victory have an effect on policy? Yes, it already has. And in years to come, I believe it will have an even greater impact."—Gary S. Becker, University of Chicago economist

"The environmentalists are for any energy source—unless it actually works."—Stephen Hayward, American Enterprise Institute



"Your Final, Final Warning. This Time We're Serious."—Rapture News, a religious news medium, forecasting the "End of Days." Echoes of these "final, final" warnings by religious zealots have been taken up by the climate change zealots of today



"For only when our arms are sufficient beyond doubt can we be certain beyond doubt that they will never be employed."—President John F. Kennedy, from his 1960 inaugural speech. Some words to ponder by those who call for the reduction of the America's military forces—especially those involving expenditures designed to keep our military the most modern and unbeatable in the world. Again we should remember that all-too-true quote "Weakness is a provocation." Implementation of the Green New Deal, which advocates extensive military cuts, will certainly result in a much weaker American nation and the military forces that protect it, making conflict with nations that feel they are militarily stronger the ultimate result.

"Americans want from government not flights of fancy but sobriety; not ecstatic evocations of dreamlike tomorrows but a tolerably functioning today; not fantasies about a world without scarcities and therefore without choices among our desires and appetites but a mature understanding of the limits to government's proper scope and actual competence."—George Will

"The Democrats' Green New Deal calls for a fully renewable electrical power grid. Regardless of the economic or political challenges of bringing this about, it is likely technologically impossible. An electric power grid requires second-by-second balancing between generated supply and consumer demand. . . This doesn't work for wind and solar because you can't spontaneously increase wind or sunshine. . . . Fossil fuel turbines, by contrast, naturally compensate for sudden supply outages . . . An all renewables power grid is destined to collapse."—Robert Blohm, member of the North American Electric Reliability Corp.,

"A zebra does not change its spots."—2000 Democratic presidential candidate Al Gore—this nonsensical quote from the leading political advocate of the dire effects of global warming.

"The average man I encounter all over the country regards government as sort of a great milk cow, with its head in the clouds eating air, and growing a full teat for everybody on earth."—Clarence Manion

"Al, the people we are going to visit are suffering. The president doesn't want to hear about your global warming crap."—Assistant to President Clinton speaking to Vice President Gore, who had implied the weather related event, whose victims the president was going to visit, may have been caused by global warming.  As related in Roy W. Spencer's book *Climate Confusion*.

"The broad mass of a nation will more easily fall victim to a big lie than to a small one."—Adolf Hitler

"Welcome to the capitalist system.  Each one of you is responsible for the amount of money you have in your pocket.  The Government is not responsible for whether you eat, or whether you're poor or rich.  The Government doesn't guarantee you a job or a house.  You've come to a rich and powerful country, but it's up to you whether or not you continue living like you did in Cuba."—Alex Alvarez, warning Cuban immigrants what they would encounter in America.

"So you can imagine how I feel when I see the U. S. making the same mistakes that Britain has made: expanding its government, regulating private commerce, centralizing its jurisdiction, breaking the link between taxation and representation, abandoning its sovereignty. You deserve better, cousins. And we expect better."—Daniel Hannan, one of Great Britain's representatives in the European Parliament.


"Arlington (Virginia) officials boast the (federal green energy) project will save $14,000 in annual electricity costs, but the solar panels have a life span of no more than 10 to 15 years. So the feds spent $300,000 to shave at most $150,000 off the net present value of Arlington's electric bills."—Stephen Moore


"At no time shall the number of employees in the Department of Agriculture exceed the number of farmers in the United States."—Proposed amendment to the law that established the Department of Agriculture, which was laughed down by Congressional representatives. It's no laughing matter today. In 1900, at the time this amendment was proposed, there were six million farms and the Agriculture Department had 9000 employees. Today, there are less than two million farms and more than 100,000 Agriculture Department employees—and the trend continues.

"America will never be destroyed from the outside. If we falter and lose our freedoms, it will be because we destroyed ourselves."—Abraham Lincoln


"(North Dakota's economy) sticks out like a diamond in a bowl of cherry pits."—Ron Wirtz, editor of *fedgazette*, commenting on the precarious financial position of most states in 2011, principally due to underfunded public pension plans, high taxes, high unemployment and poor economies. North Dakota's unemployment rate was only 3.8%, as a result of a surging economy due to oil exploration and production, moderate taxes and its right-to-work status. 650 oil wells were drilled in 2010 and another 5,500 wells are planned over the next two decades. It was one of the few northern tier states to show an increase in population after several decades of decreases.


"Natural slaves."—Aristotle, describing those citizens who would like the government to take care of them cradle to grave.

"If you are not prepared to use force to defend civilization, then be prepared to accept barbarism."—Thomas Sowell

"Save the planet, kill yourself."—Bumper sticker, mocking the true feeling of environmentalists and climatologists towards the human race.

"The resulting famines could be catastrophic."—1975 Newsweek article forecasting a looming ice age, based on the Earth's past geologic history, since the past 11,600 years of the planet's warm interglacial period is coming to an end

"It isn't pollution that's harming the environment. It's the impurities in our air and water that are doing it."—Vice President Al Gore

"Too many religious leaders have no understanding how economies work, and thus they focus on redistributing wealth without regard for how wealth is created."—Michael Novak

"The Commanding General is well aware that the forecasts are no good. However he needs them for planning purposes."—Response from a general's aide to a report by Kenneth Arrow, an Army statistician and his colleagues that the long range weather forecasts that their unit had been reporting were found to be correct only 50% of the time—a number equal to pure chance. From a report titled "Wrong Again" by Andrew Ferguson in *The Weekly Standard*.

"We have so much aluminum; it's running out of our ears."—Roosevelt administration official near the close of World War II. The Roosevelt administration chose to put the production of important wartime commodities in the hands of the country's industrialists rather than attempt to administer the economy by government decree— a decision which produced an abundance of the wartime goods needed to make America "the arsenal of democracy". Compare this experience with today's attempt by the federal government to dictate the country's healthcare system.

"No drilling or mining—not anywhere, not any time."—Unspoken goal of environmentalists.

"In general, the art of government consists of taking as much money as possible from one party of the citizens to give to the other."—Voltaire (1843) Some things never change.

"Life was nasty, brutish and short."—Unknown. This phrase has been widely used to describe the lives of the vast majority of the common people before the rise of capitalism and the industrial revolution, deemed to have started around 1820. The lives of common people  improved exponentially for those countries who embraced capitalism, including those in Europe, the United States, Canada, Australia, Singapore, Japan, South Korea, and finally, the most obvious recent example—China.

"Socialism in general has a record of failure so blatant that only an intellectual could ignore or evade it."—Thomas Sowell

"The seductive promise of security from cradle to the grave is the real enemy of civilized society."—Alexis de Tocqueville

"Democratic capitalism is neither the Kingdom of God nor without sin. Yet all other known systems of political economy are worse. Such hope as we have for alleviating poverty and for removing oppressive tyranny—perhaps our last, best hope—lies in this much-despised system."— Michael Novak, who made an intellectual journey from socialism to capitalism.

"Socialism is the residue of the Judeo-Christian faith without religion. It is a belief in the goodness of the human race and paradise on earth. Capitalism is a system built on belief in human selfishness; given checks and balances, it is nearly always a smashing, scandalous success. — Theologian Michael Novak from his commentary "A Closet Capitalist Confesses", which he wrote after many years of supporting liberal and socialist causes.

"We must deflate the pretensions of self-appointed elites. These elites will hate us no matter what we do, and it is legitimate for us to help dump them into the dustbin of history."— Eric Hoffer from his book *The True Believer*

"A few students discovered that pompous teachers who catechized them about academic free speech could, with a little shove be made into dancing bears."—Adam Bloom, from his book *The Closing of the American Mind,* describing the radical students of the 60s and 70s, who took over the campuses at that time and objected vociferously against free speech, which expressed ideas in opposition to their beliefs.  This intolerance arose again, with a vengeance in 21st century, regarding man-made climate change.

"The trouble with socialism is that it would take too  many evenings."—Oscar Wilde, alluding to the fact that the workers would have to decide what to make and what to charge, decisions currently made by the marketplace in a capitalistic society.

"Oh, progressives, if you only had the wisdom to back off, to see your demands as maximalist, extreme, damaging to the fabric, the opposite of live and let live.  When you push in this way to control the culture of the country, do you ever ask, 'When I win, will there be a country left?"—Peggy Noonan

"The (power) grid is designed to shut down, and we're designing it to shut down more often. . . .You only have to dig a quarter inch into the reliability literature to see, that while weather is always with us, and while 'major events' will tend to steal the show, renewable intermittency is the new systematic challenge to grid reliability. Renewables are a puzzle, both directly and indirectly, because they suck up investible resources that might be used for other purposes. . . . When the design performance limitations of utility systems come into play, it will always be in the interest of politicians and utility executives to change the subject to global warming."—Holman Jenkins. Jr.

"We are socialists! We are enemies of today's capitalistic system that exploits the economically weak."—Adolf Hitler

"The essential notion of a capitalistic society is voluntary cooperation, voluntary exchange. The essential notion of a socialistic society is force"—Milton Friedman

"Any fourth grade history student knows that socialism has failed in every country, at every time in history. President Obama and the Democrats are either idiots or deliberately trying to destroy their own country."—Vladimir Putin, President of Russia

The climate change people have no proof of their claims. They have computer models that do not prove anything."—David Bellamy, Conservationist

"The American people will never knowingly accept socialism. But under the name of liberalism they will accept every facet of the socialistic program, until one day America will be a socialistic nation, without knowing how it happened."—Norman Thomas, former Socialistic candidate for president

"Every one of the world's major polluting countries must institute draconian conservation measures, shut down much of its energy and transportation infrastructure and completely retool its economy. . . Human beings, including millions of government-hating Americans, need to accept high taxes and severe curtailment of their familiar life styles without revolting. They must accept the reality of climate change and have faith in the extreme measures taken to combat it. . . Every day, instead of thinking about breakfast, they have to think about death."—Novelist Jonathan Franzen. Well, now that's your pleasant thought for today from a hysterical climate change believer. I recommend he stick his head into his breakfast cereal and milk, and take a deep breath.

# APPENDIX I

## SPEECH GIVEN BY GOVERNOR NIKKI HALEY
## AT THE HUDSON INSTITUTE
## FEBRUARY 26, 2020

It's a great pleasure to be at the Hudson Institute. Hudson is in the forefront of the battle of ideas. Ideas about what makes America strong and prosperous. Ideas about what makes the world freer and more just. I'm honored to be with you.

I came here today with a simple goal. To remind us why capitalism is the best economic system the world has ever seen.

A growing number of people have forgotten this fundamental truth.

The Democratic presidential candidates are embracing socialism. And some in more traditionally conservative circles are calling for a "watered-down" version of capitalism.

While these two camps are different, they are the same in one key respect. Both are dangerous to the American people. For the sake of our children and grandchildren, America's future must be built on an actual capitalist foundation.

I'll start with the people and place I know best: My family and my home state of South Carolina.

I'm the daughter of Indian immigrants. My parents left lives of relative privilege in India to move to America. They wanted to give their children the best shot at the best life. It took them to the town of Bamberg, South Carolina, population 2,500.

When I was growing up, my mother did what so many immigrants do. She started a small business. A retail clothing and gift store in one of the most rural parts of the state. It was a family business in every sense of the word. Started from scratch.

When I was twelve years old, my mother's long-time bookkeeper announced she was leaving. After a couple of weeks of not finding a replacement, the bookkeeper got concerned. She had to train someone to take her place and time was growing short. She asked my mom how she wanted to handle it.

I happened to be walking past at that exact moment. My mom grabbed my arm and said, "Train her. She can do it."

By the time I was 13, I was doing taxes, keeping the ledger, and balancing the expenses and bank account. It wasn't until I got to college that I realized that wasn't normal.

Now I realize it was totally child labor.

I love telling this story for many reasons. Mostly because it shows the promise of America.

This is a country where a woman can start a business.

Where she can hire neighbors and give them a good paycheck.

Where people can find different jobs, better jobs, that match their talents and passions.

This is a country where someone who wants to make the world a better place can.

And for all these reasons, this country has lifted up more people, unlocked more progress, and unleashed more prosperity than any other country in history.

This is America. And the American system is capitalism.

Many people avoid saying that word, including some conservatives and business leaders. Some think it's a "politically incorrect" word. But we shouldn't be ashamed of capitalism. It's another word for freedom. And it springs from America's most cherished ideals.

We all know the most famous phrases from the Declaration of Independence. "We hold these truths to be self-evident." We are "endowed by our Creator with certain unalienable rights."

But while we know these words, we often forget their meaning. Yes, they mean free speech, freedom of religion, and other fundamental rights. But the Founders knew that economic freedom was also essential. What good are our rights if our homes and savings can be taken? Are we really free if we own nothing and the government owns everything? Of course not.

So while the Founders never used the word, they gave us capitalism in all but the name. And over the years, we've proven that capitalism can work for everyone.

It is a myth that capitalism is just for the wealthy or big corporations.

It has benefited workers and communities most of all.

It is also a myth that capitalism is all America stands for. We have other deeply held values that make our economy and our country stronger.

We don't allow sweatshop working conditions in America.

We've invested in a vast public education system to help our children learn the skills they need.

We've created a social safety net to make sure our friends, family, and neighbors don't fall through the cracks when they fall on hard times.

As Americans, we don't want a country where people get left behind.

And we don't want a country where government tells people, "You can't aspire to a better way of life. We'll just try to make you comfortable with the life you have."

We want a country where every American can get ahead.

And we have made tremendous progress.

Every American is massively better off today than our ancestors were.

In 1800, no one had electricity. If you needed to go somewhere, you probably walked. You were lucky if you lived to the age of 40. One-third of children didn't live past the age of five.

Then Americans did what we do best. We worked hard. We invented things. We gave more people the freedom to pursue their dreams. We reached for the stars. And we never let anything get in our way.

Now we drive cars. We fly in airplanes. We have the internet, computers, and cell phones. Average incomes have soared by more than 4,000 percent. Amazing medical breakthroughs mean we live twice as long. And America is just getting started.

But it's not just us. Capitalism has transformed the world.

Two hundred years ago, 94% of the world lived in extreme poverty. Today, it's 10%.

Much of this drop happened in the last 40 years, after Soviet communism collapsed in Russia and Eastern Europe, and after Communist China adopted sweeping market reforms.

It's a similar story with childhood disease, literacy, and even the environment. You wouldn't know it from listening to the gloom and doom of the left, but the facts are clear. The world is getting cleaner, healthier, and wealthier.

And the main reason is capitalism. Everywhere capitalism takes root, people do better.

We should celebrate this, not apologize for it.

I've seen capitalism from two perspectives: Private life and public life.

My family's experience taught me that capitalism can lift up families and communities. It also taught me that government can have your back, or it can stab you in the back. That lesson was a constant reminder when I got elected.

As Governor of South Carolina, I would always talk about how hard it was to make a dollar, and how easy it was for government to take it. My goal was to flip the script and come down on the side of workers and job creators. I wanted to make it clear that government was there to serve the people, not the other way around.

We started by changing the government culture. We had all state employees start answering the phones by saying, "It's a great day in South Carolina. How can I help you?" They hated it. But I wanted to remind them that they worked for the person on the other end of the phone.

I told my agencies that time is money, and if you were costing a person or business time, then you were costing them money, and that was no longer acceptable. We had every agency streamline things. We were determined to set businesses up to succeed, not regulate them into failure.

The results were incredible.

We attracted over $20 billion in new capital investment. We were number one in foreign capital investment. Number one in our region in export growth. We were building more BMWs than any place in the world. We recruited Volvo and Mercedes Benz, and five international tire companies. By the time we ended our administration, business and trade magazines named South Carolina "the beast of the southeast," which I loved!

And best of all, our state unemployment rate hit a fifteen-year low. When I left office, more South Carolinians were working than ever before.

Now, remember, this is the same South Carolina that just one generation before was devastated by the collapse of the textile industry. For over a century, textiles dominated manufacturing in our state. The process of shifting away from it was wrenching. Real people in real communities suffered.

But our answer was never socialism. The same communities in upstate South Carolina that once had textile mills today build cars, tires, and medical devices. The region has never had more jobs or better wages than it has today.

South Carolina is proof that capitalism works, and it works best for working people. I'm proud of our state and I'm proud of our record.

Then I went to the United Nations. More than anything else, that experience demonstrated to me just how exceptional America is. How fortunate we are to have the system we do.

Even with all the progress of the last few decades, there's still too much pain and poverty in the world. There are many causes. But the biggest cause of human suffering is socialism.

Socialism is the polar opposite of capitalism. It's the dangerous idea that the government should control the economy. That government should control your ideas, your property, your money, your lives. And the consequences are well-documented and heartbreaking.

Socialism has failed everywhere it's ever been tried. Instead of prosperity, it leads to poverty. Instead of opportunity, it creates dependency. Instead of liberty, it means oppression at home and often means aggression abroad. Instead of hope for a better tomorrow, socialism means a living nightmare every single day.

Look at North Korea. Its governing document is literally called the "Socialist Constitution." While I was ambassador, the UN released a report documenting that country's plight. About 70% of North Koreans require food aid. More than one out of five young and middle-aged women are malnourished. Nearly 30% of children are physically deformed from hunger. Torture and murder are the rule in the world's deadliest gulags. And North Korea's leader dreams of overtaking his southern neighbor and threatening America.

Across the border, in China, while the introduction of market capitalism has done wonders for the economy, the socialist political system has created the world's most repressive nation. Nearly 1.4 billion people are under constant surveillance. Corruption is everywhere. In Xinjiang, millions of Muslim Uighurs have been thrown into modern-day concentration camps, forced to change their names and drop their religion. Where people do have a choice, in Hong Kong and Taiwan, no one is choosing the Chinese socialist model.

Closer to home, socialism is alive — but not well — in Cuba and Nicaragua. I dealt with these countries as ambassador. My heart broke for their citizens. They live with tyranny beyond anything Americans would recognize.

Yet nothing moved me like the suffering of the good people of Venezuela.

In 2018, I stood on the Simon Bolivar Bridge, which crosses from Venezuela into Colombia. I watched thousands of Venezuelans go by. Entire families walking in the blazing heat for hours to get to Colombia where they would have the only meal they would eat that day. The average Venezuelan lost 24 pounds in 2017 alone. Four million have fled their homeland. They are literally fleeing socialism.

The socialist dictator Nicolas Maduro, propped up by his Cuban socialist allies, has run Venezuela into the ground. It was the richest country in Latin America when it was capitalist. It also had free and fair elections. Now Venezuelans are digging in trash cans and killing zoo animals for food. Millions are sick, starving, and denied the most basic political freedoms.

The same day I stood on the Simon Bolivar bridge, I met a Venezuelan family and held their beautiful baby girl. Her mom was emotional. She said she never wanted her baby to grow up that way.

Socialism is a total disaster. And as Americans, we must condemn it, wherever it exists.

That is why it is truly amazing to see how socialism has become trendy in parts of America.

These days, it seems like socialism is everywhere.

It's in our colleges and universities.

It's in Congress, where an up-and-coming Congresswoman says, "capitalism is irredeemable."

It's at the highest levels of society and politics.

Right now, the Democratic presidential frontrunner is an avowed socialist named Bernie Sanders.

Only in a prosperous country like America can people be so flippant about capitalism and so naïve about socialism. Do you know where people aren't demanding socialism? Venezuela. Nicaragua. North Korea. Every socialist country ever.

It's at this point in the argument that defenders of socialism usually say that of course they don't mean the socialism of Venezuela or China.

Their version is different, they say. It's kinder and gentler. They point to Scandinavia.

The same Scandinavia where Sweden tried socialism, saw it fail, and went so far in the other direction that it now has one of the freest economies in Europe.

The same Scandinavia where Denmark cut its business tax rate by more than half. The Danish Prime Minister criticized Bernie Sanders and said his country is a "market economy."

And get this. Finland's president was recently asked if his country was socialist. His response: "No, God bless."

Other democracies have tried socialism. Israel, India, and the United Kingdom went through periods of socialism, only to abandon it. Their people are markedly better off as a result.

This trend in America must be taken seriously. An entire generation has grown up without knowing the suffering caused by socialism in the 20th century. America's collective amnesia is becoming a real threat. We owe it to ourselves and to future generations to educate young Americans about where socialism leads.

We have the facts on our side. But more than that, we've seen the faces of socialism's victims. Their stories deserve to be heard. Stories of breadlines that stretch around streets. Stories of empty grocery stores and hospitals without medicine. Stories not just of poverty, but of oppression, torture, and murder. Just ask any of our fellow American citizens who fled socialism. They know the truth — and we need to tell it.

This is a fight we must wage, and win. We must promise each other, and future generations, that America will never become a socialist country.

But socialism isn't the only thing we need to guard against.

There's a growing trend of people who reject socialism and capitalism. They say the free market is broken, and it needs to be fixed. I find this strange, because it often comes from people who should know better.

Take the Business Roundtable. It includes the CEOs of America's biggest companies. Last year, they tried to redefine the role of business. These corporate leaders want companies to focus not on business, but on some vague notion of helping "stakeholders." They seem embarrassed by their companies' success. They shouldn't be.

What the Business Roundtable says it means by "stakeholders" is customers, workers, and communities. This is puzzling.

It's puzzling because in a capitalist economy, companies must tend to their customers, employees, and communities in order to succeed.

A company that doesn't serve its customers, doesn't have them for very long, and will go out of business.

A company that doesn't reward its workers sees them go to work for a competitor.

A company that is not a good citizen in its community breeds resentment and ultimately will not prosper.

A cutthroat company that cheats its customers, abuses its workers, and offends its community is going to fail every time.

The Business Roundtable knows this. They know what capitalism is and why it works.

The thing that allows abusive companies to survive is misguided government policies.

Remember Solyndra?

It was supposed to cure global warming and create thousands of jobs.

It did neither. And when it went bankrupt, taxpayers were left holding the bag.

Which brings us to other critics of capitalism.

Some conservatives have turned against the market system. They tell us America needs a "new kind" of capitalism. A different kind of capitalism. A hyphenated capitalism.

Yet while these critics keep the word capitalism, they lose its meaning. They want to give government more power to make more decisions for businesses and workers. They differ from the socialists only in degree.

Others reject capitalism in the name of the environment. Too many young Americans are being convinced that the only way to save the planet is to reject capitalism.

The opposite is true. Only countries that succeed economically have the resources and political will to demand a cleaner environment. Our capitalist economy continues to grow even as our carbon emissions decreased in 2019.

It falls to all of us to explain to young Americans that if they care about preserving the environment, they should care about preserving capitalism.

That is not to say our system is perfect. Of course it's not.

The critics of capitalism – whether socialists or embarrassed conservatives — have pointed out some real issues. They are just wrong when they try to diagnose the underlying disease.

They're right when they say too many businesses engage in corrupt self-dealing. We saw it in the housing crisis of the last recession. We see it today with some anti-market monopolistic behavior. But that's not capitalism. It's corruption. It's often illegal, and it's always immoral. Corruption has no place in a free market. Everyone deserves an equal shot.

They're right that too many special interests get special treatment. But that's not capitalism either. That's cronyism and corporate welfare. It destroys a level playing field and rigs the economy in favor of the well-connected. We should expose it and root it out. And no company should ever get a taxpayer bailout.

They're also right that some communities in the American heartland have suffered ill effects from globalist economics. But globalism and capitalism are not even close to synonymous. Take it from me, at the United Nations, I had a front row seat to witness the values of the multilateral bureaucrats. I assure you capitalism was not among them.

Finally, critics of capitalism are right that income levels are unequal in America. Income inequality will always exist in a free economy. That is not capitalism's proudest feature, but it's infinitely better than the alternative. Under socialism, everyone is equal. But they are equal in their poverty and misery. Most Americans don't want to imagine John Lennon's world of no possessions. We want everyone to have the opportunity to prosper, and capitalism creates that.

Our country has problems. Most of them are driven by cultural decay and big government, not by businesses or billionaires. Broken families. Insecure borders. Schools that cost too much and teach too little. A safety net that often traps people. Capitalism can't be blamed for any of these problems. In fact, it can solve them. And the past few years have proved it.

President Trump has tackled economic issues head-on, from trade to taxes to red tape. Unemployment has hit a 50-year low. Wages are up for working people. Food stamp rolls are down. Companies of all kinds have created millions of new, good-paying jobs. The stock market has created trillions in new wealth, helping millions of retirees. After decades of investing overseas, our businesses are investing in American towns and American workers.

The United States is on a roll. Because this administration brought capitalism back.

Socialists and hyphenated capitalists have no good answer to this. And if their diagnosis is wrong, then it stands to reason that their cures are wrong, too.

Their so-called "solutions" always end up with the federal government taking a bigger role in telling businesses and workers what to do, how to invest, how to live. More tax credits here. More subsidies there. More mandates for this. More regulations for that.

The underlying assumption is that they can design a better economy.

My question to them is simple. If politicians can best run the economy, then why is Washington, D.C. such a mess?

There's also another big problem with this approach.

For conservatives, the trend toward asking companies to get more involved in public issues is dangerous. If companies jump into politics, we'll get more corruption, more collusion, and more corporate welfare, not less.

And by now we should all realize just how captive the corporate world is to liberal political correctness.

Disney threatens to pull out of Georgia because of a pro-life law.

Pension funds consider boycotting Israel and bowing to anti-Semites.

Google refuses to work with the Pentagon to keep America safe, even while its search engine is freely used by terrorists.

This is what happens when companies become activists. It's rarely a good thing for conservatives. And it's bad for workers too. Few things are more dangerous than big government and big business in bed with each other. Businesses should stick to business.

At the end of the day, hyphenated capitalism is no capitalism at all. The better name for it is socialism lite. And it's just a slippery slope to the full-blown thing.

I don't believe we'll get there. Because I believe in the American people. We have proved for nearly 250 years that capitalism works.

As Americans, we cannot give up on the values that made us the envy of the world. We must put those values to work once again, as we are doing today.

And while we renew our economy at home, we must proudly promote our principles abroad. Foreign adversaries like Russia and China seize on our failure to defend our way of life.

We must respond to the critics with the truth. Capitalism can end poverty, feed the world, and cure disease. Capitalism can give every child the chance not just to dream, but to do. That's the hope that every parent has for her children.

And so, my thoughts come back to my family. My mom. My dad. My sister and two brothers. The six of us together, making our way in Bamberg, South Carolina, where our neighbors didn't know who we were, what we were, or why we were there. We were there because my parents knew that in America their children would be better off than they were. I thank God every day for their decision to come here.

Our journey was unique, but our story was not. It is the same story that generations of Americans have lived. A story of hope, hard work, hard times, and hard choices. A story of belief that tomorrow will be better than today. It was. And it still can be for this generation, and all who follow, if we make the right choices.

Speaking for myself, I will never forget the lessons I learned in my mother's store, and in all the days since. And I will never stop being the loud and proud ambassador of those principles to my country and the world.

Thank you. And God bless America.

Author's Note: Please consider supporting The Hudson Institute and Stand For America. https://www.hudson.org/about/support https://standforamericanow.revv.co

# APPENDIX II

# THE PARIS CLIMATE ACCORD

*The Parties to this Agreement,*

*Being* Parties to the United Nations Framework Convention on Climate Change, hereinafter referred to as "the Convention",

*Pursuant* to the Durban Platform for Enhanced Action established by decision 1/CP.17 of the Conference of the Parties to the Convention at its seventeenth session,

*In pursuit* of the objective of the Convention, and being guided by its principles, including the principle of equity and common but differentiated responsibilities and respective capabilities, in the light of different national circumstances,

*Recognizing* the need for an effective and progressive response to the urgent threat of climate change on the basis of the best available scientific knowledge,

*Also recognizing* the specific needs and special circumstances of developing country Parties, especially those that are particularly vulnerable to the adverse effects of climate change, as provided for in the Convention,

*Taking full account* of the specific needs and special situations of the least developed countries with regard to funding and transfer of technology,

*Recognizing* that Parties may be affected not only by climate change, but also by the impacts of the measures taken in response to it,

*Emphasizing* the intrinsic relationship that climate change actions, responses and impacts have with equitable access to sustainable development and eradication of poverty,

*Recognizing* the fundamental priority of safeguarding food security and ending hunger, and the particular vulnerabilities of food production systems to the adverse impacts of climate change,

*Taking into account* the imperatives of a just transition of the workforce and the creation of decent work and quality jobs in accordance with nationally defined development priorities,

*Acknowledging* that climate change is a common concern of humankind, Parties should, when taking action to address climate change, respect, promote and consider their respective obligations on human rights, the right to health, the rights of indigenous peoples, local communities, migrants, children, persons with disabilities and people in vulnerable situations and the right to development, as well as gender equality, empowerment of women and intergenerational equity,

*Recognizing* the importance of the conservation and enhancement, as appropriate, of sinks and reservoirs of the greenhouse gases referred to in the Convention,

*Noting* the importance of ensuring the integrity of all ecosystems, including oceans, and the protection of biodiversity, recognized by some cultures as Mother Earth, and noting the importance for some of the concept of "climate justice", when taking action to address climate change,

*Affirming* the importance of education, training, public awareness, public participation, public access to information and cooperation at all levels on the matters addressed in this Agreement,

*Recognizing* the importance of the engagements of all levels of government and various actors, in accordance with respective national legislations of Parties, in addressing climate change

*Also recognizing* that sustainable lifestyles and sustainable patterns of consumption and production, with developed country Parties taking the lead, play an important role in addressing climate change, Have agreed as follows:

## Article 1

For the purpose of this Agreement, the definitions contained in Article 1 of the Convention shall apply. In addition:

(a) "Convention" means the United Nations Framework Convention on Climate Change, adopted in New York on 9 May 1992;

(b) "Conference of the Parties" means the Conference of the Parties to the Convention;

(c) "Party" means a Party to this Agreement.

## Article 2

1. This Agreement, in enhancing the implementation of the Convention, including its objective, aims to strengthen the global response to the threat of climate change, in the context of sustainable development and efforts to eradicate poverty, including by:

(a) Holding the increase in the global average temperature to well below 2 °C above pre-industrial levels and pursuing efforts to limit the temperature increase to 1.5 °C above pre-industrial levels, recognizing that this would significantly reduce the risks and impacts of climate change;

(b) Increasing the ability to adapt to the adverse impacts of climate change and foster climate resilience and low greenhouse gas emissions development, in a manner that does not threaten food production; and

(c) Making finance flows consistent with a pathway towards low greenhouse gas emissions and climate-resilient development.

2. This Agreement will be implemented to reflect equity and the principle of common but differentiated responsibilities and respective capabilities, in the light of different national circumstances.

## Article 3

As nationally determined contributions to the global response to climate change, all Parties are to undertake and communicate ambitious efforts as defined in Articles 4, 7, 9, 10, 11 and 13 with the view to achieving the purpose of this Agreement as set out in Article 2. The efforts of all Parties will represent a progression over time, while recognizing the need to support developing country Parties for the effective implementation of this Agreement.

## Article 4

1. In order to achieve the long-term temperature goal set out in Article 2, Parties aim to reach global peaking of greenhouse gas emissions as soon as possible, recognizing that peaking will take longer for developing country Parties, and to undertake rapid reductions thereafter in accordance with best available science, so as to achieve a balance between anthropogenic emissions by sources and removals by sinks of greenhouse gases in the second half of this century, on the basis of equity, and in the context of sustainable development and efforts to eradicate poverty.

2. Each Party shall prepare, communicate and maintain successive nationally determined contributions that it intends to achieve. Parties shall pursue domestic mitigation measures, with the aim of achieving the objectives of such contributions.

3. Each Party's successive nationally determined contribution will represent a progression beyond the Party's then current nationally determined contribution and reflect its highest possible ambition, reflecting its common but differentiated responsibilities and respective capabilities, in the light of different national circumstances.

4. Developed country Parties should continue taking the lead by undertaking economywide absolute emission reduction targets. Developing country Parties should continue enhancing their mitigation efforts, and are encouraged to move over time towards economy-wide emission reduction or limitation targets in the light of different national circumstances.

5. Support shall be provided to developing country Parties for the implementation of this Article, in accordance with Articles 9, 10 and 11, recognizing that enhanced support for developing country Parties will allow for higher ambition in their actions.

6. The least developed countries and small island developing States may prepare and communicate strategies, plans and actions for low greenhouse

gas emissions development reflecting their special circumstances.

7. Mitigation co-benefits resulting from Parties' adaptation actions and/or economic diversification plans can contribute to mitigation outcomes under this Article.

8. In communicating their nationally determined contributions, all Parties shall provide the information necessary for clarity, transparency and understanding in accordance with decision 1/CP.21 and any relevant decisions of the Conference of the Parties serving as the meeting of the Parties to this Agreement.

9. Each Party shall communicate a nationally determined contribution every five years in accordance with decision 1/CP.21 and any relevant decisions of the Conference of the Parties serving as the meeting of the Parties to this Agreement and be informed by the outcomes of the global stocktake referred to in Article 14.

10. The Conference of the Parties serving as the meeting of the Parties to this Agreement shall consider common time frames for nationally determined contributions at its first session.

11. A Party may at any time adjust its existing nationally determined contribution with a view to enhancing its level of ambition, in accordance with guidance adopted by the Conference of the Parties

serving as the meeting of the Parties to this Agreement.

12. Nationally determined contributions communicated by Parties shall be recorded in a public registry maintained by the secretariat.

13. Parties shall account for their nationally determined contributions. In accounting for anthropogenic emissions and removals corresponding to their nationally determined contributions, Parties shall promote environmental integrity, transparency, accuracy, completeness, comparability and consistency, and ensure the avoidance of double counting, in accordance with guidance adopted by the Conference of the Parties serving as the meeting of the Parties to this Agreement.

14. In the context of their nationally determined contributions, when recognizing and implementing mitigation actions with respect to anthropogenic emissions and removals, Parties should take into account, as appropriate, existing methods and guidance under the Convention, in the light of the provisions of paragraph 13 of this Article.

15. Parties shall take into consideration in the implementation of this Agreement the concerns of Parties with economies most affected by the impacts of response measures, particularly developing country Parties.

16. Parties, including regional economic integration organizations and their member States, that have reached an agreement to act jointly under paragraph 2 of this Article shall notify the secretariat of the terms of that agreement, including the emission level allocated to each Party within the relevant time period, when they communicate their nationally determined contributions. The secretariat shall in turn inform the Parties and signatories to the Convention of the terms of that agreement.

17. Each party to such an agreement shall be responsible for its emission level as set out in the agreement referred to in paragraph 16 of this Article in accordance with paragraphs 13 and 14 of this Article and Articles 13 and 15.

18. If Parties acting jointly do so in the framework of, and together with, a regional economic integration organization which is itself a Party to this Agreement, each member State of that regional economic integration organization individually, and together with the regional economic integration organization, shall be responsible for its emission level as set out in the agreement communicated under paragraph 16 of this Article in accordance with paragraphs 13 and 14 of this Article and Articles 13 and 15.

19. All Parties should strive to formulate and communicate long-term low greenhouse gas emission development strategies, mindful of Article 2 taking into account their common but

differentiated responsibilities and respective capabilities, in the light of different national circumstances.

## Article 5

1. Parties should take action to conserve and enhance, as appropriate, sinks and reservoirs of greenhouse gases as referred to in Article 4, paragraph 1(d), of the Convention, including forests.

2. Parties are encouraged to take action to implement and support, including through results-based payments, the existing framework as set out in related guidance and decisions already agreed under the Convention for: policy approaches and positive incentives for activities relating to reducing emissions from deforestation and forest degradation, and the role of conservation, sustainable management of forests and enhancement of forest carbon stocks in developing countries; and alternative policy approaches, such as joint mitigation and adaptation approaches for the integral and sustainable management of forests, while reaffirming the importance of incentivizing, as appropriate, non-carbon benefits associated with such approaches.

## Article 6

1. Parties recognize that some Parties choose to pursue voluntary cooperation in the

implementation of their nationally determined contributions to allow for higher ambition in their mitigation and adaptation actions and to promote sustainable development and environmental integrity.

2. Parties shall, where engaging on a voluntary basis in cooperative approaches that involve the use of internationally transferred mitigation outcomes towards nationally determined contributions, promote sustainable development and ensure environmental integrity and transparency, including in governance, and shall apply robust accounting to ensure, inter alia, the avoidance of double counting, consistent with guidance adopted by the Conference of the Parties serving as the meeting of the Parties to this Agreement.

3. The use of internationally transferred mitigation outcomes to achieve nationally determined contributions under this Agreement shall be voluntary and authorized by participating Parties.

4. A mechanism to contribute to the mitigation of greenhouse gas emissions and support sustainable development is hereby established under the authority and guidance of the Conference of the Parties serving as the meeting of the Parties to this Agreement for use by Parties on a voluntary basis. It shall be supervised by a body designated by the Conference of the Parties

serving as the meeting of the Parties to this Agreement, and shall aim:

(a) To promote the mitigation of greenhouse gas emissions while fostering sustainable development;

(b) To incentivize and facilitate participation in the mitigation of greenhouse gas emissions by public and private entities authorized by a Party;

(c) To contribute to the reduction of emission levels in the host Party, which will benefit from mitigation activities resulting in emission reductions that can also be used by another Party to fulfil its nationally determined contribution; and

(d) To deliver an overall mitigation in global emissions.

5. Emission reductions resulting from the mechanism referred to in paragraph 4 of this Article shall not be used to demonstrate achievement of the host Party's nationally determined contribution if used by another Party to demonstrate achievement of its nationally determined contribution.

6. The Conference of the Parties serving as the meeting of the Parties to this Agreement shall ensure that a share of the proceeds from activities under the mechanism referred to in paragraph 4 of this Article is used to cover administrative expenses as well as to assist developing country

Parties that are particularly vulnerable to the adverse effects of climate change to meet the costs of adaptation.

7. The Conference of the Parties serving as the meeting of the Parties to this Agreement shall adopt rules, modalities and procedures for the mechanism referred to in paragraph 4 of this Article at its first session.

8. Parties recognize the importance of integrated, holistic and balanced non-market approaches being available to Parties to assist in the implementation of their nationally determined contributions, in the context of sustainable development and poverty eradication, in a coordinated and effective manner, including through, inter alia, mitigation, adaptation, finance, technology transfer and capacity-building, as appropriate. These approaches shall aim to:

(a) Promote mitigation and adaptation ambition;

(b) Enhance public and private sector participation in the implementation of nationally determined contributions; and

(c) Enable opportunities for coordination across instruments and relevant institutional arrangements.

9. A framework for non-market approaches to sustainable development is hereby defined to

promote the non-market approaches referred to in paragraph 8 of this Article.

## Article 7

1. Parties hereby establish the global goal on adaptation of enhancing adaptive capacity, strengthening resilience and reducing vulnerability to climate change, with a view to contributing to sustainable development and ensuring an adequate adaptation response in the context of the temperature goal referred to in Article 2.

2. Parties recognize that adaptation is a global challenge faced by all with local, subnational, national, regional and international dimensions, and that it is a key component of and makes a contribution to the long-term global response to climate change to protect people, livelihoods and ecosystems, taking into account the urgent and immediate needs of those developing country Parties that are particularly vulnerable to the adverse effects of climate change.

3. The adaptation efforts of developing country Parties shall be recognized, in accordance with the modalities to be adopted by the Conference of the Parties serving as the meeting of the Parties to this Agreement at its first session.

4. Parties recognize that the current need for adaptation is significant and that greater levels of mitigation can reduce the need for additional

adaptation efforts, and that greater adaptation needs can involve greater adaptation costs.

5. Parties acknowledge that adaptation action should follow a country-driven, gender responsive, participatory and fully transparent approach, taking into consideration vulnerable groups, communities and ecosystems, and should be based on and guided by the best available science and, as appropriate, traditional knowledge, knowledge of indigenous peoples and local knowledge systems, with a view to integrating adaptation into relevant socioeconomic and environmental policies and actions, where appropriate.

6. Parties recognize the importance of support for and international cooperation on adaptation efforts and the importance of taking into account the needs of developing country Parties, especially those that are particularly vulnerable to the adverse effects of climate change.

7. Parties should strengthen their cooperation on enhancing action on adaptation, taking into account the Cancun Adaptation Framework, including with regard to:

(a) Sharing information, good practices, experiences and lessons learned, including, as appropriate, as these relate to science, planning, policies and implementation in relation to adaptation actions;

(b) Strengthening institutional arrangements, including those under the Convention that serve this Agreement, to support the synthesis of relevant information and knowledge, and the provision of technical support and guidance to Parties;

(c) Strengthening scientific knowledge on climate, including research, systematic observation of the climate system and early warning systems, in a manner that informs climate services and supports decision-making;

(d) Assisting developing country Parties in identifying effective adaptation practices, adaptation needs, priorities, support provided and received for adaptation actions and efforts, and challenges and gaps, in a manner consistent with encouraging good practices; and

(e) Improving the effectiveness and durability of adaptation actions.

8. United Nations specialized organizations and agencies are encouraged to support the efforts of Parties to implement the actions referred to in paragraph 7 of this Article, taking into account the provisions of paragraph 5 of this Article.

9. Each Party shall, as appropriate, engage in adaptation planning processes and the implementation of actions, including the development or enhancement of relevant plans, policies and/or contributions, which may include:

(a)The implementation of adaptation actions, undertakings and/or efforts;

(b)The process to formulate and implement national adaptation plans;

(c) The assessment of climate change impacts and vulnerability, with a view to formulating nationally determined prioritized actions, taking into account vulnerable people, places and ecosystems;

(d) Monitoring and evaluating and learning from adaptation plans, policies, programmes and actions; and

(e) Building the resilience of socioeconomic and ecological systems, including through economic diversification and sustainable management of natural resources.

10. Each Party should, as appropriate, submit and update periodically an adaptation communication, which may include its priorities, implementation and support needs, plans and actions, without creating any additional burden for developing country Parties.

11. The adaptation communication referred to in paragraph 10 of this Article shall be, as appropriate, submitted and updated periodically, as a component of or in conjunction with other communications or documents, including a national adaptation plan, a nationally determined

contribution as referred to in Article 4, paragraph 2, and/or a national communication.

12. The adaptation communications referred to in paragraph 10 of this Article shall be recorded in a public registry maintained by the secretariat.

13. Continuous and enhanced international support shall be provided to developing country Parties for the implementation of paragraphs 7, 9, 10 and 11 of this Article, in accordance with the provisions of Articles 9, 10 and 11.

14. The global stocktake referred to in Article 14 shall, inter alia:

(a) Recognize adaptation efforts of developing country Parties;

(b) Enhance the implementation of adaptation action taking into account the adaptation communication referred to in paragraph 10 of this Article;

(c) Review the adequacy and effectiveness of adaptation and support provided for adaptation; and

(d) Review the overall progress made in achieving the global goal on adaptation referred to in paragraph 1 of this Article.

## Article 8

1. Parties recognize the importance of averting, minimizing and addressing loss and damage

associated with the adverse effects of climate change, including extreme weather events and slow onset events, and the role of sustainable development in reducing the risk of loss and damage.

2. The Warsaw International Mechanism for Loss and Damage associated with Climate Change Impacts shall be subject to the authority and guidance of the Conference of the Parties serving as the meeting of the Parties to this Agreement and may be enhanced and strengthened, as determined by the Conference of the Parties serving as the meeting of the Parties to this Agreement.

3. Parties should enhance understanding, action and support, including through the Warsaw International Mechanism, as appropriate, on a cooperative and facilitative basis with respect to loss and damage associated with the adverse effects of climate change.

4. Accordingly, areas of cooperation and facilitation to enhance understanding, action and support may include:

(a) Early warning systems;

(b) Emergency preparedness;

(c) Slow onset events;

(d) Events that may involve irreversible and permanent loss and damage;

(e)Comprehensive risk assessment and management;

(f) Risk insurance facilities, climate risk pooling and other insurance solutions;

(g) Non-economic losses; and

(h) Resilience of communities, livelihoods and ecosystems.

5. The Warsaw International Mechanism shall collaborate with existing bodies and expert groups under the Agreement, as well as relevant organizations and expert bodies outside the Agreement.

## Article 9

1. Developed country Parties shall provide financial resources to assist developing country Parties with respect to both mitigation and adaptation in continuation of their existing obligations under the Convention.

2. Other Parties are encouraged to provide or continue to provide such support voluntarily.

3. As part of a global effort, developed country Parties should continue to take the lead in mobilizing climate finance from a wide variety of sources, instruments and channels, noting the significant role of public funds, through a variety of actions, including supporting country-driven strategies, and taking into account the needs and priorities of developing country Parties. Such

mobilization of climate finance should represent a progression beyond previous efforts.

4. The provision of scaled-up financial resources should aim to achieve a balance between adaptation and mitigation, taking into account country-driven strategies, and the priorities and needs of developing country Parties, especially those that are particularly vulnerable to the adverse effects of climate change and have significant capacity constraints, such as the least developed countries and small island developing States, considering the need for public and grant-based resources for adaptation.

5. Developed country Parties shall biennially communicate indicative quantitative and qualitative information related to paragraphs 1 and 3 of this Article, as applicable, including, as available, projected levels of public financial resources to be provided to developing country Parties. Other Parties providing resources are encouraged to communicate biennially such information on a voluntary basis.

6. The global stocktake referred to in Article 14 shall take into account the relevant information provided by developed country Parties and/or Agreement bodies on efforts related to climate finance.

7. Developed country Parties shall provide transparent and consistent information on support for developing country Parties provided

and mobilized through public interventions biennially in accordance with the modalities, procedures and guidelines to be adopted by the Conference of the Parties serving as the meeting of the Parties to this Agreement, at its first session, as stipulated in Article 13, paragraph 13. Other Parties are encouraged to do so.

8. The Financial Mechanism of the Convention, including its operating entities, shall serve as the financial mechanism of this Agreement.

9. The institutions serving this Agreement, including the operating entities of the Financial Mechanism of the Convention, shall aim to ensure efficient access to financial resources through simplified approval procedures and enhanced readiness support for developing country Parties, in particular for the least developed countries and small island developing States, in the context of their national climate strategies and plans.

## Article 10

1. Parties share a long-term vision on the importance of fully realizing technology development and transfer in order to improve resilience to climate change and to reduce greenhouse gas emissions.

2. Parties, noting the importance of technology for the implementation of mitigation and adaptation actions under this Agreement and recognizing existing technology deployment and

dissemination efforts, shall strengthen cooperative action on technology development and transfer.

3. The Technology Mechanism established under the Convention shall serve this Agreement.

4. A technology framework is hereby established to provide overarching guidance to the work of the Technology Mechanism in promoting and facilitating enhanced action on technology development and transfer in order to support the implementation of this Agreement, in pursuit of the long-term vision referred to in paragraph 1 of this Article.

5. Accelerating, encouraging and enabling innovation is critical for an effective, long term global response to climate change and promoting economic growth and sustainable development. Such effort shall be, as appropriate, supported, including by the Technology Mechanism and, through financial means, by the Financial Mechanism of the Convention, for collaborative approaches to research and development, and facilitating access to technology, in particular for early stages of the technology cycle, to developing country Parties.

6. Support, including financial support, shall be provided to developing country Parties for the implementation of this Article, including for strengthening cooperative action on technology development and transfer at different stages of

the technology cycle, with a view to achieving a balance between support for mitigation and adaptation. The global stocktake referred to in Article 14 shall take into account available information on efforts related to support on technology development and transfer for developing country Parties.

## Article 11

1. Capacity-building under this Agreement should enhance the capacity and ability of developing country Parties, in particular countries with the least capacity, such as the least developed countries, and those that are particularly vulnerable to the adverse effects of climate change, such as small island developing States, to take effective climate change action, including, inter alia, to implement adaptation and mitigation actions, and should facilitate technology development, dissemination and deployment, access to climate finance, relevant aspects of education, training and public awareness, and the transparent, timely and accurate communication of information.

2. Capacity-building should be country-driven, based on and responsive to national needs, and foster country ownership of Parties, in particular, for developing country Parties, including at the national, subnational and local levels. Capacity-building should be guided by lessons learned, including those from capacity-building activities under the Convention, and should be an effective,

iterative process that is participatory, cross-cutting and gender-responsive.

3. All Parties should cooperate to enhance the capacity of developing country Parties to implement this Agreement. Developed country Parties should enhance support for capacity-building actions in developing country Parties.

4. All Parties enhancing the capacity of developing country Parties to implement this Agreement, including through regional, bilateral and multilateral approaches, shall regularly communicate on these actions or measures on capacity-building. Developing country Parties should regularly communicate progress made on implementing capacity-building plans, policies, actions or measures to implement this Agreement.

5. Capacity-building activities shall be enhanced through appropriate institutional arrangements to support the implementation of this Agreement, including the appropriate institutional arrangements established under the Convention that serve this Agreement. The Conference of the Parties serving as the meeting of the Parties to this Agreement shall, at its first session, consider and adopt a decision on the initial institutional arrangements for capacity-building.

## Article 12

Parties shall cooperate in taking measures, as appropriate, to enhance climate change education, training, public awareness, public participation and public access to information, recognizing the importance of these steps with respect to enhancing actions under this Agreement.

## Article 13

1. In order to build mutual trust and confidence and to promote effective implementation, an enhanced transparency framework for action and support, with built-in flexibility which takes into account Parties' different capacities and builds upon collective experience is hereby established.

2. The transparency framework shall provide flexibility in the implementation of the provisions of this Article to those developing country Parties that need it in the light of their capacities. The modalities, procedures and guidelines referred to in paragraph 13 of this Article shall reflect such flexibility.

3. The transparency framework shall build on and enhance the transparency arrangements under the Convention, recognizing the special circumstances of the least developed countries and small island developing States, and be implemented in a facilitative, non-intrusive, non-punitive manner, respectful of national sovereignty, and avoid placing undue burden on Parties.

4. The transparency arrangements under the Convention, including national communications,

biennial reports and biennial update reports, international assessment and review and international consultation and analysis, shall form part of the experience drawn upon for the development of the modalities, procedures and guidelines under paragraph 13 of this Article.

5. The purpose of the framework for transparency of action is to provide a clear understanding of climate change action in the light of the objective of the Convention as set out in its Article 2, including clarity and tracking of progress towards achieving Parties' individual nationally determined contributions under Article 4, and Parties' adaptation actions under Article 7, including good practices, priorities, needs and gaps, to inform the global stocktake under Article 14.

6. The purpose of the framework for transparency of support is to provide clarity on support provided and received by relevant individual Parties in the context of climate change actions under Articles 4, 7, 9, 10 and 11, and, to the extent possible, to provide a full overview of aggregate financial support provided, to inform the global stocktake under Article 14.

7. Each Party shall regularly provide the following information: (a) A national inventory report of anthropogenic emissions by sources and removals by sinks of greenhouse gases, prepared using good practice methodologies accepted by the Intergovernmental Panel on Climate Change and

agreed upon by the Conference of the Parties serving as the meeting of the Parties to this Agreement; and (b) Information necessary to track progress made in implementing and achieving its nationally determined contribution under Article 4.

8. Each Party should also provide information related to climate change impacts and adaptation under Article 7, as appropriate.

9. Developed country Parties shall, and other Parties that provide support should, provide information on financial, technology transfer and capacity-building support provided to developing country Parties under Articles 9, 10 and 11.

10. Developing country Parties should provide information on financial, technology transfer and capacity-building support needed and received under Articles 9, 10 and 11.

11. Information submitted by each Party under paragraphs 7 and 9 of this Article shall undergo a technical expert review, in accordance with decision 1/CP.21. For those developing country Parties that need it in the light of their capacities, the review process shall include assistance in identifying capacity-building needs. In addition, each Party shall participate in a facilitative, multilateral consideration of progress with respect to efforts under Article 9, and its respective implementation and achievement of its nationally determined contribution.

12. The technical expert review under this paragraph shall consist of a consideration of the Party's support provided, as relevant, and its implementation and achievement of its nationally determined contribution. The review shall also identify areas of improvement for the Party, and include a review of the consistency of the information with the modalities, procedures and guidelines referred to in paragraph 13 of this Article, taking into account the flexibility accorded to the Party under paragraph 2 of this Article. The review shall pay particular attention to the respective national capabilities and circumstances of developing country Parties.

13. The Conference of the Parties serving as the meeting of the Parties to this Agreement shall, at its first session, building on experience from the arrangements related to transparency under the Convention, and elaborating on the provisions in this Article, adopt common modalities, procedures and guidelines, as appropriate, for the transparency of action and support.

14. Support shall be provided to developing countries for the implementation of this Article.

15. Support shall also be provided for the building of transparency-related capacity of developing country Parties on a continuous basis.

## Article 14

1. The Conference of the Parties serving as the meeting of the Parties to this Agreement shall periodically take stock of the implementation of this Agreement to assess the collective progress towards achieving the purpose of this Agreement and its long-term goals (referred to as the "global stocktake"). It shall do so in a comprehensive and facilitative manner, considering mitigation, adaptation and the means of implementation and support, and in the light of equity and the best available science.

2. The Conference of the Parties serving as the meeting of the Parties to this Agreement shall undertake its first global stocktake in 2023 and every five years thereafter unless otherwise decided by the Conference of the Parties serving as the meeting of the Parties to this Agreement.

3. The outcome of the global stocktake shall inform Parties in updating and enhancing, in a nationally determined manner, their actions and support in accordance with the relevant provisions of this Agreement, as well as in enhancing international cooperation for climate action.

## Article 15

1. A mechanism to facilitate implementation of and promote compliance with the provisions of this Agreement is hereby established.

2. The mechanism referred to in paragraph 1 of this Article shall consist of a committee that shall be expert-based and facilitative in nature and function in a manner that is transparent, non-adversarial and non-punitive. The committee shall pay particular attention to the respective national capabilities and circumstances of Parties.

3. The committee shall operate under the modalities and procedures adopted by the Conference of the Parties serving as the meeting of the Parties to this Agreement at its first session and report annually to the Conference of the Parties serving as the meeting of the Parties to this Agreement.

## Article 16

1. The Conference of the Parties, the supreme body of the Convention, shall serve as the meeting of the Parties to this Agreement.

2. Parties to the Convention that are not Parties to this Agreement may participate as observers in the proceedings of any session of the Conference of the Parties serving as the meeting of the Parties to this Agreement. When the Conference of the Parties serves as the meeting of the Parties to this Agreement, decisions under this Agreement shall be taken only by those that are Parties to this Agreement.

3. When the Conference of the Parties serves as the meeting of the Parties to this Agreement, any

member of the Bureau of the Conference of the Parties representing a Party to the Convention but, at that time, not a Party to this Agreement, shall be replaced by an additional member to be elected by and from amongst the Parties to this Agreement.

4. The Conference of the Parties serving as the meeting of the Parties to this Agreement shall keep under regular review the implementation of this Agreement and shall make, within its mandate, the decisions necessary to promote its effective implementation. It shall perform the functions assigned to it by this Agreement and shall:

(a) Establish such subsidiary bodies as deemed necessary for the implementation of this Agreement; and

(b) Exercise such other functions as may be required for the implementation of this Agreement.

5. The rules of procedure of the Conference of the Parties and the financial procedures applied under the Convention shall be applied mutatis mutandis under this Agreement, except as may be otherwise decided by consensus by the Conference of the Parties serving as the meeting of the Parties to this Agreement.

6. The first session of the Conference of the Parties serving as the meeting of the Parties to this

Agreement shall be convened by the secretariat in conjunction with the first session of the Conference of the Parties that is scheduled after the date of entry into force of this Agreement. Subsequent ordinary sessions of the Conference of the Parties serving as the meeting of the Parties to this Agreement shall be held in conjunction with ordinary sessions of the Conference of the Parties, unless otherwise decided by the Conference of the Parties serving as the meeting of the Parties to this Agreement.

7. Extraordinary sessions of the Conference of the Parties serving as the meeting of the Parties to this Agreement shall be held at such other times as may be deemed necessary by the Conference of the Parties serving as the meeting of the Parties to this Agreement or at the written request of any Party, provided that, within six months of the request being communicated to the Parties by the secretariat, it is supported by at least one third of the Parties.

8. The United Nations and its specialized agencies and the International Atomic Energy Agency, as well as any State member thereof or observers thereto not party to the Convention, may be represented at sessions of the Conference of the Parties serving as the meeting of the Parties to this Agreement as observers. Any body or agency, whether national or international, governmental or non-governmental, which is qualified in matters covered by this Agreement and which has

informed the secretariat of its wish to be represented at a session of the Conference of the Parties serving as the meeting of the Parties to this Agreement as an observer, may be so admitted unless at least one third of the Parties present object. The admission and participation of observers shall be subject to the rules of procedure referred to in paragraph 5 of this Article.

## Article 17

1. The secretariat established by Article 8 of the Convention shall serve as the secretariat of this Agreement.

2. Article 8, paragraph 2, of the Convention on the functions of the secretariat, and Article 8, paragraph 3, of the Convention, on the arrangements made for the functioning of the secretariat, shall apply mutatis mutandis to this Agreement. The secretariat shall, in addition, exercise the functions assigned to it under this Agreement and by the Conference of the Parties serving as the meeting of the Parties to this Agreement.

## Article 18

1. The Subsidiary Body for Scientific and Technological Advice and the Subsidiary Body for Implementation established by Articles 9 and 10 of the Convention shall serve, respectively, as the Subsidiary Body for Scientific and Technological

Advice and the Subsidiary Body for Implementation of this Agreement. The provisions of the Convention relating to the functioning of these two bodies shall apply mutatis mutandis to this Agreement. Sessions of the meetings of the Subsidiary Body for Scientific and Technological Advice and the Subsidiary Body for Implementation of this Agreement shall be held in conjunction with the meetings of, respectively, the Subsidiary Body for Scientific and Technological Advice and the Subsidiary Body for Implementation of the Convention.

2. Parties to the Convention that are not Parties to this Agreement may participate as observers in the proceedings of any session of the subsidiary bodies. When the subsidiary bodies serve as the subsidiary bodies of this Agreement, decisions under this Agreement shall be taken only by those that are Parties to this Agreement.

3. When the subsidiary bodies established by Articles 9 and 10 of the Convention exercise their functions with regard to matters concerning this Agreement, any member of the bureaux of those subsidiary bodies representing a Party to the Convention but, at that time, not a Party to this Agreement, shall be replaced by an additional member to be elected by and from amongst the Parties to this Agreement.

## Article 19

1. Subsidiary bodies or other institutional arrangements established by or under the Convention, other than those referred to in this Agreement, shall serve this Agreement upon a decision of the Conference of the Parties serving as the meeting of the Parties to this Agreement. The Conference of the Parties serving as the meeting of the Parties to this Agreement shall specify the functions to be exercised by such subsidiary bodies or arrangements.

2. The Conference of the Parties serving as the meeting of the Parties to this Agreement may provide further guidance to such subsidiary bodies and institutional arrangements.

## Article 20

1. This Agreement shall be open for signature and subject to ratification, acceptance or approval by States and regional economic integration organizations that are Parties to the Convention. It shall be open for signature at the United Nations Headquarters in New York from 22 April 2016 to 21 April 2017. Thereafter, this Agreement shall be open for accession from the day following the date on which it is closed for signature. Instruments of ratification, acceptance, approval or accession shall be deposited with the Depositary.

2. Any regional economic integration organization that becomes a Party to this Agreement without

any of its member States being a Party shall be bound by all the obligations under this Agreement. In the case of regional economic integration organizations with one or more member States that are Parties to this Agreement, the organization and its member States shall decide on their respective responsibilities for the performance of their obligations under this Agreement. In such cases, the organization and the member States shall not be entitled to exercise rights under this Agreement concurrently.

3. In their instruments of ratification, acceptance, approval or accession, regional economic integration organizations shall declare the extent of their competence with respect to the matters governed by this Agreement. These organizations shall also inform the Depositary, who shall in turn inform the Parties, of any substantial modification in the extent of their competence.

## Article 21

1. This Agreement shall enter into force on the thirtieth day after the date on which at least 55 Parties to the Convention accounting in total for at least an estimated 55 per cent of the total global greenhouse gas emissions have deposited their instruments of ratification, acceptance, approval or accession.

2. Solely for the limited purpose of paragraph 1 of this Article, "total global greenhouse gas emissions" means the most up-to-date amount communicated on or before the date of adoption of this Agreement by the Parties to the Convention.

3. For each State or regional economic integration organization that ratifies, accepts or approves this Agreement or accedes thereto after the conditions set out in paragraph 1 of this Article for entry into force have been fulfilled, this Agreement shall enter into force on the thirtieth day after the date of deposit by such State or regional economic integration organization of its instrument of ratification, acceptance, approval or accession.

4. For the purposes of paragraph 1 of this Article, any instrument deposited by a regional economic integration organization shall not be counted as additional to those deposited by its member States.

## Article 22

The provisions of Article 15 of the Convention on the adoption of amendments to the Convention shall apply mutatis mutandis to this Agreement.

## Article 23

1. The provisions of Article 16 of the Convention on the adoption and amendment of annexes to the Convention shall apply mutatis mutandis to this Agreement.

2. Annexes to this Agreement shall form an integral part thereof and, unless otherwise expressly provided for, a reference to this Agreement constitutes at the same time a reference to any annexes thereto. Such annexes shall be restricted to lists, forms and any other material of a descriptive nature that is of a scientific, technical, procedural or administrative character.

## Article 24

The provisions of Article 14 of the Convention on settlement of disputes shall apply mutatis mutandis to this Agreement.

## Article 25

1. Each Party shall have one vote, except as provided for in paragraph 2 of this Article.

2. Regional economic integration organizations, in matters within their competence, shall exercise their right to vote with a number of votes equal to the number of their member States that are Parties to this Agreement. Such an organization shall not exercise its right to vote if any of its member States exercises its right, and vice versa.

## Article 26

The Secretary-General of the United Nations shall be the Depositary of this Agreement.

## Article 27

No reservations may be made to this Agreement.

## Article 28

1. At any time after three years from the date on which this Agreement has entered into force for a Party, that Party may withdraw from this Agreement by giving written notification to the Depositary.

2. Any such withdrawal shall take effect upon expiry of one year from the date of receipt by the Depositary of the notification of withdrawal, or on such later date as may be specified in the notification of withdrawal.

3. Any Party that withdraws from the Convention shall be considered as also having withdrawn from this Agreement.

## Article 29

The original of this Agreement, of which the Arabic, Chinese, English, French, Russian and Spanish texts are equally authentic, shall be deposited with the Secretary-General of the United Nations.

DONE at Paris this twelfth day of December two thousand and fifteen.

IN WITNESS WHEREOF, the undersigned, being duly authorized to that effect, have signed this Agreement.

# REFERENCE SOURCES

*BOOKS:*

Arnold, Ron and Driessen, Paul, "Cracking Big Green: To Save the Earth From the Save-the-Earth-Money-Machine", CFACT, 2014

Ball, Tim, "The Deliberate Corruption of Climate Science", Stairway Press, 2014

Berman, Bob, "The Sun's Heartbeat", Black Bay Books, 2012

Driessen, Paul, "Eco-Imperialism: Green Power, Black Death", Merrill Press, 2010

Fensin, Alan, "Liars and Demons: Climate Change Truth That Anyone Can Understand", Burlington National Inc., 2015

Gray, Vincent, "The Greenhouse Delusion: A Critique of Climate Change 2001", Multi-Science Publishing Company, 2014

Hoffer, Eric, "The True Believer", HarPeren, 1989

Idso, Craig, "Why Scientists Disagree About Global Warming: The NIPCCC Report on Scientific Consensus", The Heartland Institute, 2016

Lomborg, Bjorn, "The Skeptical Environmentalist: Measuring the Real State of the World", Cambridge University Press, 2007

Lomborg, Bjorn, "Cool It: The Skeptical Environmentalist's Guide to Global Warming", Alfred A. Knopf, Inc, 2007

Morano, Marc, The Politically Incorrect Guide to Climate Change", Regnery Publishing, 2018

Milloy, Steven J., "Scare Pollution: Why and How to Fix the EPA", 2016

Montford, A. W., The Hockey Stick Illusion", Anglo-sphere Books, 2015

Newman, Lawrence W, "The Climate Change Hoax: Pathway To Socialism", Silver Millennium Publications, Inc.. 2019

Pielke, Jr., Roger, "The Rightful Place of Science Disasters and Climate Change", Consortium for Science Policy and Outcomes", 2014

Plimer, Ian, "Heaven and Earth: Global Warming, the Missing Science, Tyler Publications., 2009

Spencer, Roy, "Climate Confusion: How Global Warming Hysteria Leads to Bad Science,

Pandering Politicians and Misguided Policies That Hurt the Poor" Encounter Books, 2010

Soloman, Lawrence, "The Deniers", Richard Vigilante Books, 2015

Spencer, Roy, "An Inconvenient Deception: How Al Gore Distorts Climate Science and Energy Policy", Amazon Digital Services, 2017

Steyn, Mark, "A Disgrace to the Profession", Stockade Books, 2015


*WEBSITES:*
https://forums.tesla.com/forums/100-reasons-why-climate-change-is-natural-and-not-manmade

https://www.youtube.com/watch?v=cVkAsPizAbU&t=157sw (Critique of Paris Accord)

https://www.youtube.com/watch?v=BQHhDxRuTkI&t=141s (Freeman Dyson—oceans not rising)

https://www.youtube.com/watch?v=zDOgWeTAas0 (Impact of solar cycles on climate)

https://www.youtube.com/watch?v=tbGZo8D5gnI (Bjorn Lomborg)

*OTHER MEDIA:*

Bedard, Paul, "$CO_2$ The 'Miracle Molecule'" Key to Feeding the World", Commentary in the *Washington Examiner*, February 26, 2019

Jenkins, Jr., Holman, "Is There a Green Rational Deal?", Commentary in *The Wall Street Journal*, March 16, 2019

Jenkins, Jr., Holman, "Texas-Style Blackouts Are The Future", Commentary in *The Wall Street Journal*, February 20-21, 2021

Milloy, Steve, "The Case For A Green 'No Deal'", Commentary in *The Wall Street Journal*, April 19, 2019

Portteus, Kevin, "The False Alert of Global Warming", Commentary in the *American Spectator*, July 19, 2013

Editorial in *The Wall Street Journal*, "Texas Spins Into The Wind", February 18, 2021

www.ingramcontent.com/pod-product-compliance
Lightning Source LLC
Chambersburg PA
CBHW031513270326
41930CB00006B/398